W9-COT-939

中國出版

The Publishing Industry in China

Robert E. Baensch
editor

The Publishing Industry in China

The Publishing Industry in China

Robert E. Baensch
editor

Transaction Publishers
New Brunswick (U.S.A.) and London (U.K.)

Second printing 2004

This book is printed on acid-free paper that meets the American National Standard for Permanence of Paper for Printed Library Materials.

Library of Congress Catalog Number: 2003048422
ISBN: 0-7658-0511-1
Printed in the United States of America

Library of Congress Cataloging-in-Publication Data

The publishing industry in China / Robert E. Baensch, editor.
p. cm.
Includes bibliographical references and index.
ISBN 0-7658-0511-1 (alk. paper)
1. Publishers and publishing—China. I. Baensch, Robert E.

Z462.P83 2003
070.5'0951—dc21

2003048422

Contents

1

Introduction

Robert E. Baensch

China is the largest country in the world and is also going through major transitions at a rapid rate. There is a strong tendency to focus on September 17, 2001, when the World Trade Organization in Geneva, Switzerland, brought more than fifteen years of negotiations to a positive conclusion and agreed on the final terms for the country to join the international trade body. However, the country's amazing economic growth started back in the late 1970s. Some of the changes we see today find their roots back in 1978 when Deng Xiaoping carried out the first of a series of economic reforms. From 1978 to 1996, more than 200 million people found their way out of desperate poverty as per capita incomes quadrupled providing purchasing power for the people. When Deng died in 1997, China was receiving up to $40 billion of foreign investments a year, which was another economic threshold for the country. More than 400 of the world's top 500 multinationals are there. These are the recent major passages for a country that has been self-contained and a country that has maintained a twentieth-century form of the great wall around its nation, culture and internal trade. In that context, the most recent WTO development is the most dramatic in that the country for the first time will be openly trading with the rest of the world according the guidelines of the World Trade Organization. A country with a 1.3 billion population achieved a compound economic growth rate of 9.4 percent according to the World Bank, 9.8 percent reported by the OECD and 9.7 reported by the China National Economic Center for the time frame of 1978 to 1998. The GNP has continued to grow at a rate of 7.5 percent in 2001 and 8.3 percent in 2002. These growth rates have a major impact on the immediate neighbors in Asia and have caused a shift in the global trading system including the United States as well as Europe. Integrating China into the global economy will also have a strong effect within the country. The communications industry in general and publishing industry specifically will play very important roles during the next phase of development. The need for information especially in the business-to-business sector and all levels of education textbooks will be directly linked to the country's development.

First, it is important to provide an overview of the huge country that has taken on difficult obligations in relation to those current members of the World Trade Organization and what obligations those members can impose over short- and long-term time frames. We need to explore a brief review of the demographics, economics, legal reforms, trade, labor, environment and technology then to understand the important role the publishing industry—the information industry—will play in China.

Demographics

In relation to a geographic area, China is somewhat smaller than the United States but in population it exceeds any other country with an estimated 1,273,11,290 people as of July 2001. China takes first place as the most populous nation. However, India with a current population of over one billion and a much higher birth rate may exceed China by the year 2035. The age segment of 0 to 14 year olds has 25.01 percent of the population, 15 to 64 years old has 67.9 percent of the total, and 65 years and over represents only 7.1 percent of the population. The population growth is estimated to be .88 percent in 2001 and life expectancy to be 69.8 years for males and 73.6 percent for females.[1] Please note chapter 6, "A Growing Children's Book Publishing Industry in China," and chapter 7, "A Study of the Chinese Young Adult Reading and Its Market," to understand the dynamics of this huge market segment.

The administrative government has identified the country as twenty-three provinces and five autonomous regions and four municipalities. The country was unified under the Qin or Chin Dynasty in 221 BC, which was replaced by the Republic on February 12, 1912. The People's Republic was established on October 1, 1949, and that day is now celebrated as a national holiday. Under Jiang Zemin's leadership, the Communist party has tried to find solutions to the problems caused by the economic boom that has not been evenly shared by the people. Factory workers and peasants, which have formed the party's traditional base, are feeling increasingly alienated by corruption, job losses, and fraying of the social safety net. The party is confronted with the challenge of ruling a market economy in transition. Workers and government officials still make up the rank and file of the Communist party. The composition of China's Communist party by occupation as of June 2002 is 45.1 percent workers and farmers, 21.3 percent government officials, state service officials, party officials, army and police, 16.4 percent retired citizens, 11.6 percent technical personnel and professionals, and the balance of 5.6 percent is represented by others.[2]

Chinese households spend a comparatively small amount of their income on rent and utilities. Daily food is the biggest single expense, which is typical for poor or developing countries. The per capita annual income in U.S. dollars was $2,490.00 in the ten major cities and only $966.00 in rural regions in 1997. It increased to $2,670.00 in the cities and declined for rural regions to $870.00 in 1999. The per capita income in the major cities increased to $3,600 in 2000 and was estimated to be $ 3,850 in 2001.[3]

Education is another major challenge for this huge and diverse country. An effort is being made to launch more new centers of higher education and professional continuing education. The country recognizes that it needs to train the floating population that is coming from the farms to work in the new industrial centers. These people need more than the functional skills to work in assembly lines. According to the China Statistical Yearbook these are the summary numbers for education in China:

	Kindergartens Pre-school	Primary Schools	Secondary Schools	Universities and Colleges
Number of Institutions	175,836	553,622	939,395	1,042
Number of Students	22.4 million	130.1 million	85.2 million	5.6 million

Chinese universities started to charge tuition in 1999 for the first time in China's history as a socialist country. Prior to this threshold year, university education was free and graduates were assigned jobs by the state. The government reduced financial support to universities, therefore forcing higher education institutions and their students to carry the increasingly large burden of paying tuition. The Ministry of Education created student loan programs to assist parents and their students. Currently, Chinese universities charge an average of $700.00 annually, which represents three-fourths of China's annual per capita income.[4] In relation to this economic change of charging tuition, Chinese universities granted 465,000 science and engineering degrees in 2001, which is almost equivalent to the United States. The publishing of textbooks is very much within the administration of the government on a national and regional scale. It was not possible to include a chapter on educational textbook publishing in this first edition of the book and transitions in this sector may make it possible to provide in-depth information in a second revised edition of *The Publishing Industry in China.*

China's New Rulers

There was a considerable shift in leadership starting in November 2002 and coming to a conclusion with appointments made to key military and government posts in March 2003. The new group of rulers will be what the Chinese refer to as the Fourth Generation. Namely, Mao's generation was the first; Deng Xiaoping's the second; and Jiang Zemin's the third generation. The Politburo Standing Committee of the Chinese Communist Party exercises the highest level of power in the country. The transition from the third to fourth generation has been a comparatively calm but intense process to carry out the shift to a new generation of leadership as a result of the 16th Party Congress held in November 2002. The seven rulers in charge of China's next phase of economic development indicated that emphasis on growth will be based on rising domestic demand and less on growth generated by exports. They realize that it is time to reduce income inequalities and care for the environment. There is a consensus amidst the Fourth Generation that the long-term strategic direction for China needs to be based on the expansion of domestic demand including raising the people's living standards. There will be a new emphasis on taking effective steps to focus on family incomes of farmers, lower- and middle-income residents of small towns beyond the ten cities that have populations of more than 10 million.[5] What is also needed is a decisive approach to control rampant corruption. Graft has eroded the party's moral authority and caused discontent among those who feel bypassed by China's economic growth. The new rulers have come to power with an average age of 62 years, much better educated, aware of the country's international role, and they know how their strong political system works, including its strengths and weaknesses. Political changes in China will accommodate more diversified interests. Until the fourth generation leaders, who came to power in the process of political diversification and market economy, become the mainstream in China's political leadership, there will most likely be a prolonged period of transition in Chinese politics as the leadership tries to adapt to change taking place at a more rapid rate.[6] Meanwhile, the central government is losing another level of control as each new sector of the economy or geographic region opens up. According to David Zweig, China's development has come "to depend on local officials and bureaucrat's ability to adapt, manipulate, or evade centrally created barriers to global transactions."[7] He makes a solid case with his academic and well-referenced work that the Chinese experience is a current dynamic example for an authoritarian state to be able to open its economy to international forces without either collapsing or becoming more democratic.

Economics

The Development Research Center of China's State Council projects that the effect of China's entry in the World Trade Organization will add .25 percent per year to the growth of world trade. Improved world trade would increase world output by almost 2 percent by the year 2010 as a result of China's membership. Most of that growth is predicted to take place in the Pacific Rim or Asia because these regional economies have a closer trade relationship. For China's economic reformers, the 2008 Olympics will be another major step forward in relating to the global arena.

China's direct investments averaged at comparatively very high levels throughout the 1990s and in 1990 the country received $2.4 billion in foreign investments. That figure grew more than ten times to reach $41.7 billion by 1996. Then it averaged $40.0 billion for the next four years including 2000. Very few developing countries have ever received even $10.0 billion in any one year. China dominates the investment arena while Southeast Asia receives a small share of the total global funds.[8]

Taiwan government reports that over $60 billion has been invested in China by 80,000 different companies over the ten-year period of

1990 to 2000. Shanghai specifically and the other four major cities have drawn over 250,000 Taiwanese to move to the mainland and work in a broad range of businesses. Another local key center is Hong Kong, which moved from a British colony status to a Special Administrative Region of China in 1997. Since then, Hong Kong's strong community of professional service companies has provided a direct link to mainland China's development including accounting, banking and legal services.[9]

At the beginning of the decade direct investment flows into China comprised only 18 percent of capital investments in the developing countries of Asia. In contrast, the Association of Southeast Asian Nations (ASEAN) countries received 61 percent. However, by the end of the decade, the proportions were reversed. Namely, China received 61 percent of the investment flows and the ASEAN countries only 17 percent.[10] According to the latest statistics from the Ministry of Foreign Trade and Economic Cooperation (MOFTEC), 411,495 foreign companies had been authorized by the government to operate in China by the end of August 2002. Contracted foreign capital was about $807.6 billion, of which about $430.0 billion, was currently in use according to the Ministry. Data shows that foreign capital is playing an increasing role in China's national economy. Taxes paid by foreign-funded businesses represent 19.0 percent of the national total, while industrial added value accounts for 24.6 percent, exports value 50.1 percent and net increase of foreign exchange reserves 73.6 percent. "It is in the combination of Taiwan, Hong Kong and now rapidly developing mainland China that the future is being shaped by the best and the brightest of all three regions."[11]

Legal Reforms

The legal system has developed over time as a consolidation of a synthesis of traditions, statutes, and basic civil codes, with an emphasis on criminal law. There is a Supreme People's Court with judges appointed by the National People's Congress. Then there is a system of local People's Courts, which comprise higher, intermediate and local courts. Finally, there are Special People's Courts, which are primarily military, maritime and railway transport courts.

A new legal code has been in effect since January 1, 1980. More recently there has been a special effort to revise and develop civil, criminal and commercial law based on a major effort to train thousands of legal administrators in Europe and the United States. In

1999, Article V of the Chinese constitution was amended to include the new text that states, "The People's Republic of China shall practice ruling the country according to law, and shall construct a socialist rule-of-law state."[12]

For years the respect for and implementation of strong regulations to control and protect intellectual property has been a major problem for international companies distributing products or services in China. The International Intellectual Property Alliance estimates that piracy in China costs foreign companies an estimated $2 billion a year, or about a quarter of the total global losses attributed to copyright violations. The dramatic growth of indigenous book publishing, record, film and video industries has brought about a new perspective. Anyone who prints books, produces recorded music, makes movies or develops software recognizes that success will generate immediate piracy—copies made available in the market for a fraction of the official editions. According to conservative estimates, more than 85 percent of the videos, books, music, and software are illegal copies sold almost immediately after the original has been recognized by the market. In 1994, the U.S. Embassy in cooperation with the General Administration of the Press and Publications organized a seminar held at the Beijing Normal University for publishers to focus on "Translation and Reprint Rights." However, at that time it was still perceived as the developed wealthy countries pursuing the poorer developing countries. Since 1994, it has become more than a trade dispute because the victims are now the Chinese as reported by Joseph Kahn in the *New York Times*.[13] China's creative industries have been hit hard by the failure to enforce copyright laws. Local objections raised by local artists and film producers especially in reference to DVD copies of films and CD copies have generated more publicity than foreign legal pressure. Therefore, the local legal administration within the country has pledged to meet the international standards for protecting intellectual property. It is another test of China's weak legal system and what will be necessary to participate in the WTO's international regulations. The World Bank has accumulated relevant data that documents a direct relationship of countries with effective legal institutions are well along the path of development including the ability to feed and educate their populations. World Bank president James Wolfensohn has stated that governments need to recognize that an "effective legal and judicial system is not a luxury, but a central component of a well-functioning state and an essential ingredient in long-term development."[14] According to the WTO agreement, China promises to apply and administer its laws in a uniform, impartial and reasonable manner in implementing its regulations. In the spring of 2000 Chinese negotiators presented to the working party a list of 177 domestic laws dealing with customs administration, foreign investment administration, intellectual property, and services that required revisions to be consistent with the new international obligations of the WTO. The positive news is that China has begun the process of amending domestic laws to make its legal system consistent with WTO obligations. The results of these changes in the legal system will not occur as rapidly as Western companies and governments expect. Furthermore, China is entering a period of major transition in leadership, which could be cause for considerable concern as little is known about the individuals who are expected to assume top leadership positions in the party and government in the spring of 2003. The legal system is still primitive and acts more as a way to transmit state directives than to dispense justice in an impartial fashion. The problem is that it will take years to achieve fair, uniform and impartial implementation of the laws and to ensure that judgments are enforced.[15] The positive aspect is that the publishing industry represented by the Law Press China which publishes over 400 new titles a year and the People's Court Press are just two of several imprints that release authoritative books that Chinese judges, lawyers and international legal firms are able to consult as the legal system develops over the coming years.

Labor

Studies forecast that China will undergo significant restructuring costs of the labor force in meeting the broad WTO commitments. Measured on a purchasing power parity, China holds the position of the second largest economy in the world after the United States. Agricultural output doubled in the 1980s and industry posted major gains, helped by the above-mentioned foreign investments. One of the challenges for the huge country in transition is the floating population of 90 to 120 million rural workers who are adrift between the small villages and the ten major cities. It is a labor force surviving on part-time jobs, short-term construction work and the government's spending on building an infrastructure such as water systems, highways and the huge Yangtze river dam. The Chinese government will need enough resources to restructure the economy's industrial structure and simultaneously solve the problem of unemployment caused by the reform.[16] If China's economy does not grow by 7 percent, it will not be able to absorb the surplus labor and productivity will suffer. Therefore, it is very important that the economy maintain a high growth rate. Without such a momentum of a rapidly growing economy to absorb surplus labor, the risk of social instability will increase and reform efforts will be limited. Long term the WTO membership will help bring balance to the supply and demand for labor. In the short term, the problem is only going to intensify because in the next few years the challenge of increased unemployment of 4 to 5 percent will have to be addressed. The problem of inadequate employment in rural areas has now become a problem of unemployment in urban areas. At the same time, the skill levels for jobs have increased with related requirements for adult education and specialized training.[17] Jobs are becoming a rare commodity on a global scale and, therefore, competition for jobs is going to be more intense in China. Continued and professional training will become increasingly important as the labor force needs to move up into the skilled levels of productivity. There will become a larger demand for publishers to provide textbooks and training materials for the young adult generation that may not have had the equivalent of high school education to prepare them for jobs that demand a minimum level of technology and sophisticated assembly line production competence. Please note chapter 11, "Education of the Publishing Industry: Challenges and Developments," for a review of just what is happening within this sector of the industry to prepare the next generation to effectively manage the book, magazine and online publishing industry.

Foreign capital has fueled China's economy with more advanced and applicable manufacturing techniques. The demand has caused more highly qualified personnel to be trained and jobs created. Again, there is a demand for the publishing industry to provide the textbooks and professional information for a massive and timely upgrading of the labor force. About 23 million people are directly employed in foreign companies and thus represent more than 10 percent of the total labor force in urban areas. China's fastest growing exports have been labor intensive manufactured goods including textiles, apparel, footwear, toys and more recently all aspects of technology.

Technology

The total number of cell or mobile phone users has exploded over the past five years from less than 12 million in 1997 to reach 144.5 million by the end of 2001, thus surpassing the United States. The new total is projected by China's Ministry of Information Technology to be more than 248 million by the end of 2003. Market penetration is somewhat low for China with 11.2 percent in comparison with the developed Western Europe reaching 75 percent, North America at 44 percent. There is still considerable growth potential for this mobile phone technology in the country that has achieved the number one position internationally.

Internet users grew at a slow rate from 1997 through 1999 and then rapidly increased to reach a projected level of more than 50 million users by the end of 2002 according to the China Internet Network Information Center. To put these numbers into context, it is interesting to compare telephone lines, home PCs and Internet subscribers in the five leading countries of Asia at the end of 1999.

Country	Population (mill)	Telephone lines	Home PCs	Internet Subscribers
China	1,249.6	135,200,000	12,000,000	7,000,000
India	997.5	199,000,000	3,200,000	400,000
Japan	126.5	58,000,000	22,680,000	7,000,000
Philippines	76.8	3,000,000	825,000	240,000
Thailand	61.7	9,150,000	1,500,000	500,000

Source: International Online Markets 2000, Simba Information Inc., www.simbanet.com

The market for personal computers in China is expected to pass the Japanese market in sales in 2003 and become the world's second largest consumer of PCs after the United States. This dramatic growth has to be put into the context of purchasing power because the cost of even a basic desktop PC can equal two or three years' savings for the typical urban Chinese household despite intense price competition.[18] A focus on the major languages that will be used on the Internet provides a different view of what role China will take in the next three years. According to eMarketer studies, confirmed by Global Reach data (www.Glreach.com/globstats), the major languages on the Internet will shift as noted in the following table.[19]

Language	Year 2002	Year 2003	2003 Rank
English	45.0 %	29.0 %	1
Japanese	9.8 %	7.3 %	4
Chinese	8.4 %	20.2 %	2
German	6.2 %	5.8 %	5
Korean	4.7 %	4.4 %	6
Spanish	4.5 %	7.6 %	3

Sources: Global Reach Data, June 2001, www.Glreach.com/globstats

Motivated by the rapidly developing potential of the Chinese market and segments therein, multinationals are shifting their R&D efforts to China. For example, General Electric started a large basic research center in Shanghai, which will also assist GE's procurement of Chinese manufactured products. Intel operates five laboratories, some of which have moved beyond semiconductor research. Matsushita, Japan's largest consumer electronics company, has started a five-year research program with a budget of $330 million for an R&D center that will employ 1,500 engineers by 2005. Microsoft is working with the State Development and Planning Commission and will spend $750 million over three years on research, training, and outsourcing in China.[20] Please see chapter 5, "Scientific, Technical, Medical and Professional Publishing," for a good review of how the publishing industry is responding to the development of technology in the country by publishing both original works as well as translations for the rapidly growing market.

Environment

The environment is a challenge for this country as China has a high rate of air pollution due to greenhouse gases, sulfur dioxide particulates from their reliance on coal heating, which produces acid rain. There are water shortages, particularly in the north, and water pollution from untreated wastes. Deforestation has created major damage especially prevalent along the Yangtze River, which has turned into a dark brown flow of water due to heavy erosion. There is an estimated loss of one-fifth of agricultural land since 1949 due to soil erosion and economic development. Eighty percent of China's rivers are so polluted that they no longer support fish. The problems have reached high enough critical levels to be a threat to continued economic development even with or because of WTO membership.[21]

Air pollution, soil erosion and the continued fall of water levels in the northern regions are all issues that Wen Jiaboa, premier of the State Council, wants to address. He has moved to the forefront in taking a slower more planned approach to address air and water pollution, conservation rather than transfer of water, and deterioration of the environment by immediate government programs to exploit interior regions' resources.

Over the past ten years, Shanghai has become the largest construction site in the world, representing US$39 billion investment. Over 17 percent of the world's high-rise building cranes are found in this city. At the Shanghai Urban Planning Exhibition Center there is a huge 200:1 scale model of the city, as it will appear in 2020. The vision is an endless range of skyscrapers representing "a builder's dream and an environmentalist's nightmare."[22]

Publishing Industry

The brief overview of China and a highlight survey of the demographics, government, economics, legal reforms, technology and environment provide the challenges and opportunities that the publishing industry will face in coming years. Publishing is the *information* industry including books, magazines, professional journals and online information services. A country that is growing rapidly with major changes taking place provides a broad spectrum of demands for information and publications. New and revised textbooks need to be published for the more than 130.1 million primary school students. Or at the other end of the scale, there are the more than 480 million adults in the professional and technical labor force who will have to go back to school for continuing and professional training and thus will require texts and reference books as well as online information services.

On November 7, 1986, Gayle Feldman reported in *Publishers Weekly* that more than 1,000 publishing houses from thirty-five countries displayed 50,000 of their titles in the first Beijing International Book Fair during the week of September 5-11, 1986. She stated, "only a month earlier, many publishers wondered if the chaos would prevail, and 24 hours before opening ceremony, books were still arriving and stands were still being set up. The event gave Western visitors a good idea of the promise, as well as the problems, of China's book world."[23] It is important to put this first international event into context with what has happened during the past years, including the dramatic opening of the market for both regional domestic and international publishers.

The General Administration of Press and Publications (GAPP) of the People's Republic of China is responsible for and directs all activities of the 566 registered publishing houses. This government agency is also responsible for the administration of publishing and distributing magazines and newspapers throughout the country. Furthermore, it is the central office that administers and assigns the ISBN (International Standard Book Numbers) and ISSN (International Standard Serial Numbers) to all new publications in China. The main office of GAPP is located in Beijing and directly administrates the 221 publishing houses located in that city. There are regional GAPP offices in the major cities responsible for the other 345 publishing houses. Some of these, such as the China Agriculture Publishing House, China Science and Technology Publishing House, China Taxation Press or The Electric Power Publishing House, have a dual reporting relationship in that they belong to government ministries. The majority of the publishing houses were started by the government after 1949, at which time they identified each to be responsible for a specific subject/market segment. This system was extended to the provinces where local publishing houses were authorized to focus on economics, science, textbooks, children's books, and a People's Publishing House which would be similar to a "trade" publisher except they are not allowed to publish fiction. University presses are the only publishing organizations that function outside of these guidelines in that they can publish academic research and reference books or journals as well as textbooks across a wide range of subject matter.

The following chapters will provide detailed information on the major sectors of magazine and book publishing. Chapter 3, "Guidelines for Magazine Publishers in China," documents the rapid changes that have taken place and how several international publishers have been able to launch local Chinese-language editions of major consumer magazines in the country. Chapter 5 addresses the specific market segments of scientific, technical, medical, and professional publishing. In view of the special opportunities and challenges presented by a huge age group, or 25.1 percent of China's population, chapters 6 and 7 focus on children's books and a study of young adult readers. The last three chapters address the functional topics of distribution and bookselling, the economics of the industry, translation rights and education of the publishing industry. The goal of this volume is to provide a practical resource to the publishing industry, which is going through major changes. It is not possible to provide a comprehensive "handbook." However, there is more than enough information to establish a working knowledge and make it possible to participate in the dynamic growth of the country through export, licensing, joint ventures or partnerships with the right publisher or distributor in the most appropriate city. Contributor's names with addresses are provided in chapter 12 for your reference and exploration to take the first steps in relating to a huge country that is on fast-forward.

Notes

1. The World Factbook, www.odci.gov/cia/publication/factbook/geos/ch.
2. New China News Agency, Communist Party Demographics, Beijing, August 12, 2002, and Economist Intelligence Unit, Country Data and Country Indicators, New York, The Economist, www.eiu.com/data.
3. Edward W. Knappman and Xiao-bin Ji, *Facts about China*, Bronx, NY: W. W. Wilson, 2002.
4. Jiang Xuequjn, In China, Student Loans May be Expanding Societal Gaps, *Chronicle of Higher Education*, December 7, 2001.
5. Andrew J. Nathan and Grace Gilley, China's New Rulers, *The New York Review*, October 10, 2002, pp. 28-32.
6. Doo-bok Park, Fitfully, Change Comes, *Joong Ang Ilbo Newspaper,* Seoul, South Korea, November 23, 2002.
7. David Zweig, Internationalizing China: Domestic Interests and Global Linkages, Ithaca, NY: Cornell University Press, 2002.
8. William B. Gamble, Investing in China, Westport, CT: Praeger, Greenwood Publishing Group, 2002.
9. Nicholas R. Lardy, *Integrating China into the Global Economy*, Washington, D.C.: Brookings Institute Press, 2002, pp. 134-136.
10. Amy Louise Kazmin, Hugh Williamson, and Sheila McNulty, Foreign Investors Desert South-East Asia for China, *Financial Times,* October 13, 2000, p. 8.
11. Dexter Roberts et. al., Greater China, the new math, *Business Week*, December 9, 2002, pp. 51-59.
12. Ronald Dworkin, Taking Rights Seriously in Beijing, *The New York Review*, November 2002, p. 17.
13. Joseph Kahn, The Pinch of Piracy Wakes China Up on Copyright Issues, *New York Times*, November 1, 2002, pp. C1-5.
14. Supachai Panitchpakdi and Mark L. Clifford, China and the WTO, Singapore: John Wiley & Sons (Asia) Pte. Ltd., 2002, p. 147.
15. Stanley B. Lubman, *Bird in a Cage: Legal Reform in China after Mao*, Stanford, CA: Stanford University Press, 1999.
16. Frontline, China in the Red: Four years, ten people and a nation in the midst of an epic transition, PBS Television, 2003.
17. Liaoang Zhoukan, China's Unavoidable Limitations, Beijing, China, *Outlook Weekly,* November 11, 2002.

18. Keith Bradsher, Chinese Computer Maker Plans Push Overseas, *New York Times, Business Day*, February 22, 2003.
19. EMarketer, eGlobal: Demographics and Usage, January 2002, New York, eMarketer Inc., 2002.
20. Bruce Einhorn, Ben Elgin, Cliff Edwards, and Linda Himelstein, High Tech in China, *Business Week,* October 28, 2002, pp. 80-88.
21. Mark Hertsgaard, *Earth Odyssey*, New York: Broadway Books, 1999, pp. 246-251.
22. Brodie Fenlon, Leashing the Economic Dragon, Toronto, Canada, *The Toronto Sun,* November 29, 2002.
23. Gayle Feldman, Beijing's First Book Fair, *Publishers Weekly*, November 7, 1986, pp. 24-29.

References

Armitage, Catherine. "The Gray Helmsman." *The Australian* (Sydney) (November 16, 2002).

Bradsher, Keith. "Chinese Computer Maker Plans Push Overseas." *New York Times*, Business Day (February 22, 2003).

Clifford, Mark L., and Supachai Panitchpakdi. *China and the WTO*. Singapore: John Wiley & Sons (Asia), Pte. Ltd., 2002.

Economist Intelligence Unit, Country Data and Country Indicators, New York, *The Economist*, www.eiu.com/data.

Einhorn, Bruce, Ben Elgin, Cliff Edwards, and Linda Himelstein. "High Tech in China." *Business Week* (October 28, 2002): 80-88.

Fenlon, Brody. "Leashing the Economic Dragon." *The Toronto Sun* (Canada) (November 29, 2002).

Frontline. "China in the Red: Four Years, Ten People and a Nation in the Midst of an Epic Transition." PBS Television, 2003.

Gamble, William B. *Investing in China*. Westport, CT: Praeger, Greenwood Publishing Group, 2003.

Hertsgaard, Mark. *Earth Odyssey, Around the World in Search of our Environmental Future*. New York: Broadway Books, 1999.

Knappman, Edward W., and Xiao-bin Ji. *Facts about China*. Bronx, NY: W. W. Wilson Inc., 2003.

Lardy, Nicholas. *Integrating China into the Global Economy*. Washington, D.C.: The Brookings Institution Press, 2002.

Liaowang Zhoukan. *China's Unavoidable Limitations. Outlook Weekly* (Beijing) (November 11, 2002).

McDougall, Colina (ed.). *Trading with China: A Practical Guide*. Maidenhead, United Kingdom: McGraw-Hill Book Company (UK) Ltd., 1990.

National Bureau of Statistics. *China Statistical Yearbook 2001*. Beijing, China: Statistics Press, 2002.

Oksenberg, Michael, Pitman B. Potter, and William B. Abnett. *Advancing Intellectual Property Rights: Information Technologies and the Course of Economic Development in China*. Washington, D.C.: National Bureau of Asian Research Analysis, vol. 7 (November 1996).

Park Doo-Bok. "Fitfully, Change Comes." *Joong Ang Ilbo* (Seoul, Korea), (November 23, 2002).

Roberts, Dexter, Mark L. Clifford, and Bruce Einhorn. "Greater China." *Business Week* (New York), (December 9, 2002): 50-58.

Sheff, David. *China Dawn: The Story of Technology and Business Revolution*. New York: HarperCollins Publishers Inc., 2002.

Spence, Jonathan. *The Search for Modern China.* New York: W. W. Norton, 1990.
Studwell, Joseph. *The China Dream.* New York: Grove Press, 2002.
U.S. Trade Representative. *United States and China Reach Accord on Protection of Intellectual Property Rights, Market Access.* Washington, DC, February 26, 1995 (www.ustr.gov/releases/1995/02/95-12—March 16, 2000).
World Bank. *China 2020: Development Challenges in the New Century.* Washington, D.C.: The World Bank, 1997.
World Bank. *China, Weathering the Storm and Learning the Lessons.* Washington, D.C.: The World Bank, 1999.
Zweig, David. *Internationalizing China: Domestic Interests and Global Linkages.* Ithaca, NY: Cornell University Press, 2002.

2

Magazine Publishing in China

Zhang Bohai

To understand recent developments and future trends, it is important to review some historical stages of China's magazine industry. In the twentieth century, that is, in the past 100 years, the development of the magazine industry shows three historical characteristics. First, China's social activities rose and fell alternately in the twentieth century. In the first half-century, China carried out anti-Qing dynasty, anti-warlord, anti-Japanese aggression and anti-Kuomintang (KMT) struggles. In the second half-century, China established a new social system and carried out political, economic, cultural and educational reforms. Under these social circumstances, China's magazine industry showed much stronger efforts to document or report on the political activities and was conscious of its social mission. This main characteristic remains until now.

The second historical phase or in the early half of the twentieth century, China's economy nearly collapsed. All walks of life, including the publishing industry, were in real difficulties. In the second half of the twentieth century, China's economy began to recover and developed slowly and tortuously. After the open policy and reforms were carried out, China's economy became quite successful in many sectors. But until now China's economic level has not exceeded that of developing countries. Under these economic circumstances, it is impossible for the Chinese magazine industry to develop to the advanced levels of Western countries, and it will not be able to form the big-scale magazine publishing groups that represent the prosperous magazine industry. China is far from the advanced level of the magazine industry in the Western world.

Translation provided by Zhang Zhuoran, Center for Publishing, School of Continuing and Professional Education, New York University, New York.

The third phase includes the last half of the twentieth century, during which China experienced the changing process from a planned economy to a market economy. The magazine publishing industry reflects this dramatic shift. From 1950 to 1970, Chinese magazines did not depend on a market-driven economy and magazine publishers did not care about profit. There was not the Western sense of bottom-line performance because the publishers were protected by China's planned economy. As a result of moving to a new market-driven economy, the ability of publishers to respond and adapt to the market was required, but it was a difficult process. From 1980, the open policy and reform spurred Chinese publishers to realize their weaknesses. The Chinese magazine industry still has not completed the transition from the planned system to market competition. It is very hard for publishers to develop strong financial management in relation to the business of circulation and advertising in the new market system. Irregularity in publishing serial publications, inexperienced business management and economic weaknesses are still serious problems that cannot be ignored in China's magazine industry.

Trends of the Magazine Industry in Recent Years

There are about 9,000 magazine titles published every year in China. Half are consumer or social science periodicals and half are scientific, technology, and professional periodicals. The total number of Chinese magazines distributed in 1997 and again in 1998 was 2.5 billion units, and that number grew to 2.96 billion in 2002. The annual gross sales of the Chinese magazine industry cannot be calculated accurately because there is no effective means of gathering statistics. It is possible to provide an estimate based on total units or copies distributed and magazines' average prices. Gross sales in 2002 were 12 billion RMB, but after a 40 percent distribution fee rate was deducted, the real gross sales volume was about 7.2 billion RMB. The advertising income in 2002 was 1,521 million RMB, of which 136 million RMB was the tax. So the sum of the sales income and advertising income was 8.585 billion RMB, about U.S. $1,946 billion. Compared with the total revenues of the magazine industry in the United States, which was U.S. $26.1 billion (announced by the American Periodical Association) in 2001,

China's magazine industry is only one-twenty-fifth of that of the United States.

The Chinese magazines with circulation or distribution of more than 1 million are no more than twenty-five titles. A small number of titles have a distribution of 5 million copies and most of the magazines have a circulation of 1 million to 3 million copies. Some of the magazines depend on sales within the market economy, but some still depend on planned economy distribution models.

Subscription is the main distribution channel in China and the national post offices take full responsibility including payment processing. Only several best-seller magazines can be bought at newsstands. Although the distribution ratio between subscription and newsstand has not been measured accurately, the latter is much smaller than the former.

Four Trends for the Future of the Magazine Industry

Size Scale Trend

Chinese publishers have a common understanding that a certain level of consolidation will be required. The magazine industry should turn decentralized, small-scale and fragile publishing houses into consolidated magazine groups with a sense of developing viable business clusters. The best way will be to encourage those famous and large magazine publishers to publish the small and currently subsidized magazines or to acquire weak magazines so as to establish magazine groups. It may also be possible to combine magazines, newspaper, books and audiovisual publishing companies into comprehensive media groups. Now these changes are fulfilled step-by-step usually in response to an economic or operational crisis. Because of the special social system in China, this process will be very slow and will encounter difficulties at several levels.

High Quality Trend

At the present time, magazine readers in China have very high requirements for editorial or content quality, so the market of high-level magazines is very promising. The most important point for high quality magazines is to have quality, relevant and wonderful content to reflect new ideas and new concepts of cultural and social standards of a country going through rapid changes. The second point includes perfect design, high technological printing and good paper to deliver a quality publication. It is estimated that more and more four-color magazines will appear in the next two or three years.

Professional Trend

Chinese readers are beginning to lose interest in general consumer magazines. They are becoming more and more interested in professional magazines, which have their own characteristics. In recent years, new magazines that focus on certain audiences, certain field or certain topics have been started with rapid readers' acceptance. Therefore, many of the older magazines have started to also define their focus or establish special columns or features in their issues.

Digital Trend

In recent years, most magazine publishing houses have made the change to use the digital method to capture keystrokes, edit text, and manage digital content from author to prepress printing plate production. They are seriously exploring and in some cases managing the change from print on paper delivery to content transferred to CD-ROM or the Internet through digital technology. In 2001, the Chinese magazine industry held an "E-magazine Design Competition" to encourage publishers to create "the fourth media magazine." Facing the coming of the digital age, Chinese publishers are exploring different models for online delivery. However, it is similar to other parts of the world where it has not been possible to recognize the concrete economic benefits of e-magazines in the consumer market segments.

Problems in China's Magazine Industry

Circulation sales via newsstands and other retail outlets are not yet fully developed because the distribution channels are not developed as commercial business enterprises. Therefore, the market of consumer magazines has not developed as quickly or as largely as it should have over the past three years. Currently, the total number of magazines distributed is 2.96 billion copies. That is to say one person in China purchases only two copies of magazines per year, but in Japan, one Japanese will purchase more than twenty copies of magazines. It is obvious that the potential market for magazines still has a very large potential for development.

The total distribution of magazines was 2.5 billion units in the 1980s. The amount did not change for more than ten years. In the last ten years, China has changed in every aspect of the social and economic development but the sale and distribution of magazines remains stagnant. The potential for the magazine industry remains untapped. The main reason for slow development is that the distribution channels for newsstand and other retail outlets have not been fully developed. A key cause for this problem is the lack of competent and experienced management of distribution in the publishing industry. The magazine publishers do not have market research, reader's demographic data and circulation databases. As a result, most publishers still do not know who their audiences are, how old they are, what their professions are. The post offices carry out 70 percent of the total distribution for publishers in China. Because of this dependence on the post offices, magazine publishers have not explored nor developed much wider ranges of distribution methods.

The advertising revenue is very small at this time. The management and sales of advertising needs to be explored, developed, and then to become a major factor in the economic growth of magazines. In recent years, the advertising industry has developed rapidly, but advertising in magazines is still in the early developmental phase. In 2002, the advertising income of magazines was 1,521 million RMB, only 1.68 percent in the total media advertising income. The advertising revenue in 2000 rose to 1134 million RMB.The advertising income on TV was 23.1 billion RMB, or 25.58 percent of the total. Poor designs and irregular operation cripple the potential and hinder the development of advertising revenues for magazines. Only recently have the magazine publishers realized that they need advertising income to survive and thus they need to develop their in-house competence to attract and sell advertising.

International Communication and Cooperation

Magazines have existed for only two or three hundred years. They comprise a comparatively new medium that appeared with the social development of China. Cultural openness, information and communication are the main characteristics not only of modern society but also of magazines. In 1815, the first Chinese magazine appeared in Melaka, which is Malaysia today. Western missionaries wanted to promote their religious doctrine and Western culture through a Chinese magazine. Therefore, promotion of Western culture and developed technology became the main characteristics of the Chinese magazines that emerged at that time. *Good Friends*, an illustrated magazine, was published in 1926. This publication had bilingual captions for the illustrations so that foreigners also could read it. *World Knowledge and Translation*, which was first published in 1934, is very famous for introducing international information and cultural development. Chinese magazine publishers intended to develop the business abroad very early. Since China established the open policy, magazines have become a more active means of promoting communication within the country. In the last twenty years, Chinese publishers had friendly relations with publishers of Britain, France, Germany, Japan, Israel, Korea, and the United States. Chinese magazines that are exported to other countries continue to increase in numbers. In 1999, the amount of exported magazines was 1.68 million issues or units. A series of Chinese magazine exhibitions was held in Europe, America, Asia, and Australia. The publishers have developed a comparatively stronger working knowledge of the cooperative ventures, international rights and especially the sale and purchase of copyrighted content. The China Periodical Association became a member of the International Federation of Periodical Press by the end of 2000. Now that China is a member of the World Trade Organization, China will be able to develop more contacts with foreign countries. More and more opportunities of cooperation will appear. Several principles that the Chinese magazine industry should follow are:

- Cooperation should develop on the base of mutual benefit and mutual respect.
- Chinese publishers will not allow foreigners to direct or dominate the Chinese market.
- Cooperative business ventures of the magazine publishing industry should respond to China's politics, economy, culture and social customs.
- International cooperation of publishers should abide by the laws and regulations in China.

China not only wants to know and understand foreign cultures but it also wishes to promote its own culture to foreign countries.

As we look into the near future, the market for magazines is the same as other markets for consumer products in China. It is very complex, large and full of potential. On the one hand, Chinese publishers are familiar with the market. On the other hand, it is hard for Chinese publishers to master the changes taking place at an increasingly rapid rate. For foreign publishers, it is even harder and more complex. How China responds to the pull of the future while also trying to implement the push of the present will be interesting to observe.

3

Guidelines for Magazine Publishing in China

Charles Buckwalter

As China's commercialization moves forward, so does its demand for the latest technical, professional, and product information. At every level of industrial development the need intensifies for China to keep aware of ideas and innovations being introduced competitively around the globe. The existing sources for much of this information include the joint-venture partnerships with major international corporations, the associations, the academic institutions, the multitude of government bureaus, as well as the carefully controlled internal publishing industry.

But as China continues to widen its role in international trade, logic dictates that the dynamics will require a greater reliance on outside information resources—particularly for the kind of technical and trade information that feeds the mechanics of a growing economic engine. The key, and most-abundant source for that information is the world's trade press in the United States, Europe, and Asia-Pacific.

Ironically, while other Chinese industries have benefited from the influx of foreign capital, technology, and management, Chinese magazine publishers are struggling to keep up with the pace of change, and are unable to access foreign or even new domestic Chinese capital due to highly restrictive rules and regulations. Advertising expenditures continue to expand, but broadcast media have done a better job of capitalizing on this growth than magazine publishers.

Although the Chinese Communist Party, through various ministry-operated structures, continues to exert rigid control over editorial content of any publication, vertical trade or business magazines

---

This chapter is based on a White Paper prepared by the International Committee of the American Business Media Association.

are less likely candidates for ideological conflict than broad-based consumer magazines. The editorial content of trade or business magazines is more-readily categorized as educational, instructional, or useful in serving the development of specific segments of Chinese industry. They have a better chance of adhering to China's guiding principal that science, technology, and education are primary productive forces in rejuvenating the nation.

Even with an optimum editorial category, publishers seeking entry into the China market must be ready to accept and operate within strange new rules that govern not only editorial content, but also every other aspect of publishing. Magazine publishing in China is virtually a new industry, evolving in a climate where political, social, and cultural forces often clash and entangle as the process moves forward. It often is, indeed, a Chinese puzzle.

In recent times, international foreign publishers more and more are seeking a share of China's lucrative, new frontier, notwithstanding the difficulties and frustrations it poses. Quoted in *Folio* magazine, Thomas D. Gorman, a leading publishing authority in the market and Hong Kong-based chairman of CCI Asia-Pacific Ltd., licensees of Time Inc.'s *Fortune* in China, states, "If you have the right outlook, do your homework carefully, and especially if you take a long-term view, there is tremendous opportunity here…[but] assumptions about what worked or was in demand three years ago, for example, may be seriously out-of-date today. This is an extremely dynamic market."[1]

Growth of the Internet. The latest phenomenon to impact the media/communications environment of China is the explosive growth of Internet users from 300,000 in 1995 to more than 50 million today. The government was originally wary but now realizes this is a huge engine for economic growth and education, and one which is essentially beyond the level of the kind of control they would like to exercise. Foreign capital is plentiful, as is capital from private and corporate sources from within China.

Largest economy. Most reliable sources agree that by the year 2010 China will become the world's largest economy, creating new industries and wide-ranging new business opportunities on the international scene. Driving this impressive growth has been the PRC's determined shift toward a more market-oriented economy implemented in stages of five-year plans over the past twenty years. The business climate for China continues to improve even in the face of tighter government regulations.

The key market regions. Although China has a population of 1.3 billion, only about 25 percent to 33 percent have viable potential as a consumer market above the basic necessities. That is China's urban middle class, which has real spending power and is growing rapidly in size. This segment of the population is found in China's three richest areas: *the Pearl River Delta* (especially the three cities of Shenzhen, Guangzhou, and Zhuhai), *the Yangtze River Delta* (principally Shanghai, and reaching to Nanjing in the north and Ningbo in the South), and *the Beijing-Tianjin corridor.* Other important economic centers include the Northeast between the port City of Dalian and Shenyang, Shandong, on the Fujian coast, and the inland centers of Wuhan, Xian, and Chengdu. Many of China's leading firms are located in the three main economic regions.

Shanghai is China's largest port with $63.64 billion worth of shipments passing through. It has a population of over 16 million and is the industrial, commercial, and financial capital, producing 5 percent of China's total industrial output. Shanghai's main industries include automobiles, textiles, telecommunications, large-scale machinery, electronics, petrochemicals, iron and steel. Shanghai also has one of China's two stock exchanges with 477 listed stocks and a 1998 trade volume of about $150 billion.

A large part of China's foreign direct investment is made in Shanghai. Almost 19,000 enterprises were contracted to invest $34.3 billion in Shanghai by the end of 1998. The United States topped the list of Shanghai foreign investors with $861 million at about 600 U.S. companies in Shanghai.[2]

The Magazine Publishing Industry

Participation in China's publishing industry by foreign companies is still severely restricted, mainly to licensing ventures, but many foreign magazines have found creative ways to enter the market. And there are clear indications of a more-positive ambience on the horizon.

China's membership in the World Trade Organization was expected to bring with it notable publishing reforms for both domestic and international players. But, so far, that has not been quite the case. Gorman explains it this way:

> For a variety of reasons the reality has been that a few things have changed on the domestic scene, but very little has changed, policy-wise, for existing or would-be international entrants into the market.

> Publishing—unlike other businesses such as insurance, telecommunications, agriculture, automotive, etc.—was not directly a part of China's WTO accession agreement; so any changes to date and in the future will be only indirectly related to the overall WTO-related reform process. So, the Chinese government has not promised anything—unlike the other industries covered in the WTO deal.

Some things that have changed are the options for domestic Chinese publishers for acquiring capital. There are more options, although they are still limited and tightly controlled. Acquiring capital from international sources is still prohibited, strictly speaking, although there are numerous examples of "creative" deals that have been done.

The big change is that Chinese publishers have been finally and officially encouraged through new policy directives to focus on the bottom line, acquire modern management techniques, and grow their businesses. It is a big change for them, but at a fairly theoretical level. Other changes include some opening up on the retail and distribution front, and some relaxation about foreign investment in the printing sector.

Gorman comments,

> Given the substantial gray area, however, most of the foreign parties to investment deals into Chinese magazines are smaller, privately owned companies versus the larger public media players whose legal departments are more likely to "choke" on the confusion, complexity, and inconsistency of the regulatory environment.

Although China's role in WTO may not provide major publishing reforms in the near term, it does set a new tone for improving management skills, increased emphasis on developing electronic media, greater cooperation in aspects of intellectual property rights, and other open-market orientation—all of which can enhance the long-term picture for quality business and trade magazines to serve China's new information needs

But for now, foreign magazines must continue to operate within the current tight structure. They can enter the market through indirect access by collaborations with Chinese organizations, ministries, agencies or government-related trade associations. Another approach is that of producing the magazine off the China mainland—chiefly in Hong Kong—and importing it. By translating, editing, and printing the magazine offshore, publishers can avoid the rule that requires one-third of editorial content to be provided by the local Chinese nationals participating in mainland publishing affiliations with foreign publishers. However, imported magazines must meet other stringent controls, such as limiting paid distribution to the handful of state-approved publications importers. Foreign publishers need to evaluate the pros and cons of every approach to find which best suites their magazine title and the target audience it hopes to reach.

Rules of the Game

The number one rule to learn about magazine publishing in China today is that there are lots of rules. Complex rules. Rigid rules. Frustrating rules. Inconsistent rules from one place to another, with inconsistent precedents and changing interpretations. What's legal and not legal is subject to constant debate and discussion. Transparency is extremely limited.

Unlike the Western world where you devise a smart publishing business plan and implement it through a fairly routine series of steps, the path to success in China is laden with stumbling blocks. One expert, Ann Stevenson-Yang, writing in *The China Business Review*[3] a few years ago put it this way:

> After foreign publishers first read China's regulations and meet with the international cooperation sections of the relevant ministries, they are likely to be thoroughly discouraged. Beijing has yet to craft a law on publishing, but apparently plans to have one in place by 2010. In the absence of such a law, the sector is guided by PRC regulations that state that no private organization or individual is permitted to own or operate a periodical or book publishing house.
>
> The chief rules governing the industry, the 1997 Regulations on Publication Management, the 1997 Regulations on Management of the Printing Industry, and the 1998 Regulations on Electronic Publishing, stipulate that only government-designated publishing organizations may engage in publishing. Thus no private individuals or foreign entities may engage directly in publishing in China. The 1995 and 1998 guiding catalogues for foreign investment in industry indicate that publishing is off-limits to foreign investment. Foreign firms are also prohibited from owning distribution companies or taking the majority share in advertising joint ventures, unless the Chinese partner is a private advertising company. Publications that are permitted to accept advertising must have an advertising agent that is registered and licensed in China.

It is now widely believed that new laws on publishing may be ready by 2003. These rules and regulations are expected to affect domestic players more than international. They will reaffirm the predominant role of the Party with regard to content issues and reassert the high fences around foreign direct investment into print media.

A dismal picture? Only if you expect to enter the China market by applying the same well-organized publishing patterns you may be accustomed to in the United States or in other international markets. In China you must rely on affiliations with Chinese partners and their capacity to help you operate within the framework of governing bodies and ministries and bureaus. You will require stamina to deal with regulations that may seem arbitrary, illogical, and can change without warning; and you will need patience to plod through the process of approvals and, ultimately, to see profitable results.

The Regulatory Bodies

Three powerful, interrelated government divisions control publishing in China. They are known to operate in disharmony and they often limit each other's ability to deal with situations that require flexibility, although the reporting structure is more clear-cut than that which governs the fast-growing Internet sector. The three bureaucracies are as follow: (1) *The General Administration for Press and Publications* (GAPP) is responsible for supervising publishing in China. It functions under (2) *The State Council* and (3) *The Publicity Department (aka Propaganda Department) of the Communist Party of China* (CPC), which is involved on key policy and project approvals. In addition to these three principal controlling groups, local branches of the Public Security Bureau (PSB) and the State Administration for Industry and Commerce (SAIC) also play supporting roles—the latter with respect primarily to the advertising business. Other PRC government agencies may also become involved where the ventures include electronic media, such as the newly established State Council Information Office and the Ministry of Culture (MOC).

The GAPP's section for international cooperation is sometimes the first contact foreign media companies have with China's publishing bureaucracy. The GAPP accepts applications for new publications from Chinese organizations, assigns publication codes and numbers and approves changes to existing publications. The SPPA also serves as the government authority that takes action against illegal publications and publishing units.

The Communist Party of China, theoretically, is not allowed to interfere directly with activities of the PRC government agencies. But in practice the CPC Publicity Department has an important hand in policy matters and in approving key projects. It also is the ultimate watchdog over editorial content for all media operating in China.

The State Council Information Office (actually linked to the CPC Publicity Department) operates directly under the State Council (China's cabinet) and it also weighs in on regulating foreign media operations. It proposes related policies; coordinates international publicity activities by the media; convenes press conferences for cabinet departments; publishes official white papers on China's official views; helps supervise resident foreign news offices; organizes news-gathering trips by foreign media; and it also regulates the Internet media.

The PRC government agencies involved in the licensing process vary depending on the type of publication activity. License approvals for magazines are generally issued by GAPP at the central level and by both GAPP and PSB (Public Security Bureau) at the local branches. Strict guidelines remain in effect, although in recent years local authorities have been known to relax them somewhat in allowing new local magazines to emerge.[4]

Identifying Markets

Identifying a particular market's business or technical information needs in China calls for more time and diligence than foreign publishers are accustomed to in the United States or other open societies. Industry market profiles, details and data sought from official Chinese bureaucratic sources may not be easily available or reliable. Independent market surveys and research can be conducted with the help of a reputable in-country partner, but publishers should be careful and skeptical in having the research conducted and in checking the results.

Excellent general market analysis is available through the U.S. & Foreign Commercial Service, the U.S.-China Business Council, The American Chamber of Commerce in China—and other U.S. and China organizations that monitor China's market growth phenomenon. One of the recent helpful documents issued by the U.S. & Foreign Commercial Services is "Contact China."[5] It includes a concise analysis of twelve principal markets to watch in China. These markets, in a broad sense, represent targets of opportunity for business publishers whose editorial products might adapt to the technical, scientific, managerial or other information needs of potential readers in the multiple specialties and talents that underlie the markets. Some of these categories may provide easier access for publishers to enter the market because they fit one of China's stated national goals of increasing education in science and technology. Having the right editorial focus is essential. But the key question for

U.S. business publishers is whether these or other markets can be tapped profitably. The answer to that is impacted by many other factors.

Here are the twelve markets listed in the "Contact China" report, followed by a brief synopsis of each:

- Aerospace and Aircraft Industries
- Automotive Industry
- Computers and Networking
- Environmental Technologies
- Health Care: Products and Services
- Health Care: Pharmaceuticals
- Housing Industry
- Power Sector
- Retail Industry
- Semiconductors and Software
- Telecommunications Equipment
- Transportation Infrastructure Industry

Excerpts *from the Market Briefs Reported in "Contact China"*

Aerospace and Aircraft Industries. China's aviation sector continues to grow faster than that of the rest of the world (at about 6 percent from 1996 to1998). Feeder and charter airlines are purchasing small numbers of 3- to 40-passenger fixed wing and rotor aircraft for both scheduled and charter flights. Purchasers of these aircraft favor U.S. and European suppliers over Russian and domestic manufacturers. Airport construction has slowed but still provides opportunities for air traffic control ground equipment and architecture/ engineering providers.

Although most of China's air space is still controlled by the military, small holes are opening up. Tourist helicopter flights over Shanghai and the Great Wall that were not available last year are now a reality. Offshore oil exploration groups have purchased helicopters. One non-airline company owns a corporate jet and more will follow suit when the amount of red tape associated with non-scheduled flights is reduced. The trend is clearly towards more open skies, making small aircraft sales a greater possibility in the coming years.

Automotive Industry. The automotive industry in China has taken a dramatic turn in the past decade. While individual purchases used to represent less than 1 percent of auto sales, they now represent about 25 percent of sales for the leading joint venture producer in China. Taxi and other companies along with government bureaus round out the remaining 75 percent. Most autos in China used to be easier to build, maintain, and service, but now Chinese cars are more stylish and energy-efficient, to meet the changing needs of more discriminating consumers. Although previously expanding, the overall auto market is now at a virtual standstill at approximately 511,000 units per year.

Computers and Networking. Computing has entered a network era and China has felt the impact. Related information technology sectors have witnessed rapid growth and many new digital products are entering the China market. Though domestic firms make competitive products, foreign involvement in technology and new ventures is expanding. With increases in power and speed and concurrent decreases in price, PCs are now within the reach of individual Chinese buyers. U.S. computers still hold a small market share, but Chinese computers with U.S. components dominate the market. PCs are becoming essential appliances for Chinese families who want the best for their children, who have unprecedented buying power due to "4-2-1" demographics (4 grandparents, 2 parents, and 1 child). Information appliances linked to Internet access are poised to attack the market at even lower prices. The growth of subscribers and users of the Internet is stunning: by July 1999 at least 6.7 million subscriptions shared by multiple users pointed to many more users than frequently reported.

Environmental Technologies. China's acute environmental problems stem from a deteriorating natural resource base, dense population, heavy reliance on soft coal, outmoded technology, under-priced water and energy, and breakneck industrial growth. The World Bank estimates that air and water pollution cost the Chinese economy up to 8 percent of GDP. In response, the government has unleashed a burst of environmental legislation, but local enforcement is spotty, investment in pollution control infrastructure inadequate, and competition from domestic firms increasingly strong. Products enjoying the best sales prospects include low-cost flue gas desulfurization systems, air and water monitoring instruments, drinking water purification products, vehicle emissions control and inspection devices, industrial wastewater treatment equipment, and resource recovery technologies.

Health Care: Products and Services. Over the next few years, China's health care sector will be characterized by uneven growth for U.S. exporters of health care products and services. Profit margins in the approximately $14 billion medicines market are expected to fall in [the short term] due to recent government cost containment measures. Prospects for dietary supplements and medical and dental equipment continue to look promising, but financing remains a key concern. U.S. and foreign health care service providers, attracted by the large expatriate population and China's small but growing middle class, are finding ways to enter the market.

Health Care: Pharmaceuticals. Until recently, China's pharmaceutical market has been one of the fastest growing markets in the world. From 1990 to 1996, the pharmaceutical market experienced over 200 percent growth and has now expanded into a $14 billion market. While overall demand will continue to grow at 10 percent, import and joint-venture product market share and profits are expected to fall. Joint-venture drugs account for 50 to 60 percent of the drug market. In 1998, the value of imported drugs was around $1 billion, but due to pressures from the reimbursement system, which favors domestic medicines, the import drug market share is gradually shrinking. In 1998, gross output value reached $19 billion, increasing 14 percent over the previous year. The domestic industry is characterized by non-branded generic production, overproduction, and losses. The government is consolidating China's 6,000 pharmaceutical enterprises, of which 71 percent are state—and collectively—owned.

Housing Industry. During President Clinton's visit to China in July 1998, the Presidential Housing Initiative was presented, paving the way for U.S.-China cooperation on housing issues in both the construction and financing arenas. Since that time, a Housing Council has been formed with the Ministry of Construction (MOC) and the Departments of Commerce and Housing and Urban Development (HUD). This Council will help educate builders, regulators, and consumers on the more high tech end of the building materials market as well as housing financing options.

The Chinese government has recently stated growth rates for housing in the 20 to 25 percent range, down from earlier estimates of 35 percent. Although this is probably still too high, more conservative figures suggest there will be growth in housing of about 15 percent for the next few years.

Power Sector. China's power industry, while enjoying rapid growth through most of the 1980s has recently experienced a significant slowdown in demand and a commensurate decline in its growth rate. The drop is mainly attributable to an overall decrease in the consumption of coal by 14 percent. Coal, which accounts for three-quarters of China's primary energy, is not alone. The consumption of oil, gas and nuclear fuel was also down. Hydropower, usually comprising 5 to 6 percent of China's basic energy needs, probably grew. As a result of this contraction in the demand for energy, China has undertaken to slow the growth of its base of installed electric generating capacity.

China will increase the number of new hydroelectric power projects and will undertake refurbishment of older, inefficient plants. Much of this retrofitting will concentrate on environmental technologies designed to combat China's chronic pollution problems. U.S. and foreign companies still enjoy clear superiority in more technically advanced systems that control electrical power stations and the transmission and distribution of power. Market conditions favor the installation of smaller, efficient gas turbines to provide peak-load, cogeneration or in off-grid or "inside the fence" configurations.

Retail Industry. China's retail sales rose by more than 25 percent per year between 1995 and 1997. This explosive growth rate slowed in 1998 due to the Asian economic crisis and its bigger-than-anticipated impact on China's economy. Total retail spending in 1998 was $351.2 billion.

Positively predisposed toward foreign-brand products, Chinese consumers increasingly have the means and desire to purchase. Leasing international brand-name consumer goods are capturing dominant market positions as they displace lower-quality brands. Chinese manufacturers are becoming increasingly more brand and quality conscious as well, as evidenced by the fact that local brands have the dominant share of PC and consumer appliance markets.

As manufacturers vie for the attention of the Chinese consumers, the advertising industry is experiencing a record boom. From a base of zero in 1979, it rose to annual 1998 billings of $6.5 billion, making china the world's 7th largest market.

Semiconductors and Software. Firms related to information technologies such as semiconductors, co-processors, and software have expanded into China, but the international dominance of American firms does not yet extend to a growing China market. However, the rapid development of microprocessors and network software means that these new markets are offering potential for growth.

Chinese manufacturers currently face a lack of technical know-how. China supports the entrance of foreign firms into the semiconductor and components manufacturing markets, but continuing problems with intellectual property rights and difficulties in creating entirely owned foreign ventures or foreign-controlled joint ventures represent significant barriers to entry. Over 90 percent of U.S. electronic components used in China are purchased in Hong Kong and brought across the border in Guangdong.

Telecommunications Equipment. Since the early 1990s, China's telecommunications market has grown about 10 percent per year, yet nationwide teledensity reaches only about 12 percent. China planned to invest about $20 billion in its telecommunications sector in 1999.

The separation of the Ministry of Information Industry (MII) and China Telecom—at both the central and provincial levels—into regulator and operator, respectively, was scheduled to be completed by the end of 1999. The breakup of China Telecom into fixed-line, mobile, satellite, and paging entities, the strengthening of China Unicom (China's second carrier) with the transfer of Guoxin Paging Company, and the formation of additional operators is *expected to increase competition.*

Transportation Infrastructure Industry. The Chinese government is in the midst of a massive upgrade to existing transportation systems. For example, China added 52,000 km of highway in 1998 alone. The number of bridges, rail lines, stations and airports has also increased. This trend is expected to continue with projects in highways, railways, and ports, and include opportunities in construction equipment, engineering, and electronics and safety devices.

According to World Bank statistics, goods lost due to poor and obsolete transportation infrastructure amounted to 1 percent of China's GDP as recently as the most current survey (mid-1990s). Foreign companies have been banned from engaging in freight forwarding unless they form a joint venture with local partners. It is hoped that with accession to the WTO these and other structural issues will be resolved, enabling U.S. firms to compete, and add value to China's infrastructure system.

Realities of Magazine Publishing in China

Unlike many international markets, foreign publishers who do business in China must depend to some extent on Chinese partners and the Chinese bureaucratic systems for operating profitably. Although China is lumbering steadily toward a more market-oriented economy, it still operates with ideological reservations about the free flow of ideas. The Chinese government's primary concern with the entry of new magazines and other print information is the social and political impact—not the economic benefit. China is not the place for publishers with little or no international experience to seek quick profits. Even well-established publishers who are operating profitably find the process an ongoing challenge year-to-year.

Finance and Payments

Finance is a key question because China's currency, the Renminbi, is non-convertible into U.S. dollars or other hard currencies. One common arrangement for licensing deals for international advertising is that the foreign publisher sells to, bills, and collects from international companies in U.S. dollars or other hard currencies, and relies on this revenue stream to support its share of income from the deal. Most Chinese publishing companies do not have access to U.S. dollars or other hard currency, so relying on these for payments of royalties or other fees is generally not viable or practical. For this reason, American publishers must specifically address the question of payment and remittance arrangements with potential Chinese counterparts, publishing partners or licensees as part of the business feasibility study phase. Renminbi income is only useful to U.S publishers with sufficient operations in China to justify payment of local operating or travel costs in Renminbi. The same issue affects the sale of advertising in China to local Chinese companies, some of which have easy access to U.S. dollars for payment, and some of which do not.

Editorial Content

One of the first realities foreign publishers face in approaching the China market is that of editorial control. For business and trade media—due to their vertical, technical nature—the problem is apt to be less severe and easier to manage than broad-based consumer magazines. How a magazine's editorial content will be impacted depends largely on how the publication establishes itself for distribution to potential readers in the country. The two possibilities are:

1. *Establishing a joint venture or other collaborative agreement* with the publishing arm of an official Chinese or quasi-official Chinese organization. This might be a ministry or bureau in charge of promulgating a particular trade or special interest, a government-affiliated trade association, or the like. In this arrangement, the foreign publisher gets access to targeted audiences through the internal lists controlled and maintained under the Chinese partner's auspices. This is a plus, especially for controlled circulation magazines that need a well-defined reader base and access to demographics. However, this arrangement impacts editorial control. Under Chinese law, *no foreign individual or company may control editorial content and at least one-third of the editorial content of any internal publication with foreign involvement must be locally generated.* The possible consequence for foreign trade magazines that pride themselves on the quality of their editorial content is that of producing a Chinese edition with a third of its editorial mix not up to Western standards.

   But there is also a potential upside to working with such an editorial mix. That is the adaptability factor. Advanced technical editorial matter published in leading trade magazines of the Western world could often be unsuitable or impractical for application by readers in comparable Chinese functions. Editorial content, to be kept useful and adaptable, might very well benefit by having a one-third input from the local Chinese side.

   The editorial–advertising mix is also a factor. Apart from other restrictions, Chinese domestic magazines are limited by regulation to no more than a 30/70 advertising/editorial ratio.

   Under these publishing arrangements, the rules are strict. Chinese regulations state that such ventures can publish "sound content," but the definition of what's "sound" is left vague, meaning that the PRC government can openly scrutinize foreign publishing companies. And penalties for violations are stiff, including fines, revocation of publisher's license—even criminal charges.

2. *Importing magazines printed offshore*. Foreign publishers can avoid the requirement regarding one-third local editorial content by opting to have their magazine printed wholly offshore and imported to the China mainland, rather than establish an inside joint venture. The import set-up is common for foreign publishers, especially mass-circulation consumer magazines. It does protect editorial quality. But to make it work effectively it still require a solid, inside partner who can

deal with the special permissions and navigate the regulations that apply. One difficulty of going the import route is that all of these magazines must be distributed through one of the handful of state-approved publications importers, which does not give access to most newsstands. That regulation poses particular problems for trade magazines that need to develop a sufficient reach among well-defined reading audiences. Editorial content may be "100 percent pure" yet the magazine may not be readily available to those who need it most.

The Language

Although China still has many local spoken dialects of the Chinese language, the PRC long ago adopted Mandarin Chinese (Putonghua) as the standard national spoken dialect. Fortunately for publishers, written Chinese is and always has been standard and uniform, irrespective of local differences in spoken dialect (unlike India, for example, which has more than 200 different languages that differ in both spoken and written form). The main variation, which applies to written and printed Chinese characters, is that the PRC adopted a simplified form of Chinese characters to promote broader literacy. "Simplified Chinese" is an abbreviated script in which each character has less strokes than in the "traditional" or complex script. Simplified Chinese is the standard form in the PRC, whereas in Taiwan and Hong Kong the traditional or complex script is still used. For newspaper literacy, a reading knowledge of about 3,000 Chinese characters is the generally accepted norm. Apart from the difference between simplified and traditional Chinese characters, there are significant differences in writing style and terminology between the PRC, Hong Kong, and Taiwan (somewhat akin to, but greater than the differences between British and American English, for example), so it pays to ensure that for PRC audiences editors are experienced in producing content for the PRC rather than Hong Kong and Taiwan audiences.

Plagiarism

Abuses in the area of Intellectual Property Rights are rampant in China. Even though China has extensive IPR legislation on the books, in practice, ensuring enforcement and promoting education regarding these rights is a long, slow process. It is not uncommon for local Chinese publishers to reproduce articles appearing in their U.S. counterpart magazines without asking permission or giving any acknowledgement whatsoever. The copyright holder may have little recourse in protecting his rights due to the tedious, complex nature of the legal system. Chinese courts are just beginning to cope with the situation.

Circulation

All domestic Chinese magazines circulated in China are tracked by the assignment of a unique periodical number. That number is issued by China's State Press and Publications Administration (SPPA). The number is assigned based on the magazine's specific characteristics, including number of pages, frequency of publication, and other factors. Once these characteristics are approved and the number assigned, they may not be changed without approval of GAPP. Only one number applies to one magazine. It may not be used for multiple titles. Getting approval for a publication number is increasingly difficult. China's government is currently taking steps to *reduce* the quantity of active publication permit numbers it has issued, due to a proliferation of periodicals and books on the market.

Foreign publishers who do not have joint venture affiliations with internal Chinese entities, do not need a periodical number for producing their magazine offshore and importing it into China through China Books and Periodicals Import and Export (I/E) Corporation, or through China Educations Publications I/E Corp. Imported magazines fall into three control categories: (1) Business Science and Technical publications; (2) General Interest publications; and (3) Reactionary publications. The business, scientific, and technical magazines are easiest to import.

There is but one publications niche for which a publication number is not required. That is publications such as newsletters or daily briefs put out by association or government organizations. To avoid the need for a number, these publications must be registered as having circulation strictly limited to their group. Legally, they are prohibited from accepting advertising and being distributed outside the group. In reality, some of these publications, which are sponsored by powerful inside governmental operations, do accept advertising and sell copies openly to anyone interested in the subject matter. They are sometimes challenged by regulatory agencies depending on the degree to which they tend to serve the general educational interest and that the publication content does not conflict with national political philosophy. Publications of this kind have been used as "shells" by both foreign and domestic publishers to channel specialized magazines to specific markets. But there is always a risk in choosing this route. By working in collaboration with the sponsoring organization, foreign publishers have been able to totally redesign the content and design of these publications (leaving just enough graphic "identity" to keep them qualified), circulate them effectively through the system, and run them effectively as shared commercial vehicles.[6]

Circulation Auditing Services

Controlled circulation magazines have been in China since the 1970s, but are primarily published offshore by foreign publishers. *BPA* now counts about 22 PRC publications among its members, mostly produced in or for China by or in cooperation with international publishers. *ABC* audits several magazines in Hong Kong, including *Time Asia* and *Newsweek Asia*, and it has begun auditing newspapers within the China mainland through its Hong Kong office.

Distribution

Magazine distribution in China is complicated to say the least. Both paid circulation and controlled circulation magazines suffer problems.

Distribution of mass-market magazines is achieved largely by the Chinese Post Office, which has 60,000 outlets nationwide, as well as by the largest bookseller, Xinhua (New China) bookstore. There are over 13,000 state-owned bookstores and bookstands, largely dominated by Xinhua. The largest Xinhua bookstore is located in the Xidan District of Beijing. There are an even greater number of private ones (collectively owned). Foreign companies are not allowed to own publication distribution companies in China.

Distribution of paid subscription magazines through the mail is handled by the Chinese Post Office. The post office not only delivers, it collects payments in advance directly from subscribers, charging publishers an average of 40 percent of the cover price. The publisher does not get his share of the revenue until after copies are delivered to the subscriber. Subscriber lists are controlled by the post office. The magazine publisher is given total numbers, but no demographics or details about the subscribers. Such information is considered secret.

Some Chinese publishing firms have recently been finding innovative ways to smooth the distribution problem. They are setting up their own distribution networks as a special service for customers. One such system uses bicycle delivery routes to hand-deliver magazines to subscribers. The system can assure prompt delivery, collect payments, and gather basic reader profile data. Although not widespread, techniques like this are a sign of progress.[7]

Lists and Databases

Finding good, clean circulation target lists that penetrate the China market is a tough challenge for foreign business and trade magazine publishers. Here again, strong in-country alliances are imperative. Foreign publishers can approach list building through mutual cooperation with Chinese trade associations or government ministries seeking to develop the technical/professional interests of the groups they serve. However, even when such groups might allow distribution of publications to their lists, the names and addresses are likely to be shielded and not provided to foreign collaborators for databases.

Other possible sources are international corporations that have already established a presence in China and have built up their own target lists. Such companies might be open to list-sharing if their marketing objectives would benefit from the media's introduction. Lists may also be available through traditional sources and techniques, such as: agencies of international development and cooperation, including the embassies, the U.S. & Foreign Commercial Service, the U.S.-China Business Council, the Chambers of Commerce and others. Also valuable is attendance and involvement with professional congresses, trade show, seminars, and trade missions organized by either Chinese or U.S. government departments.

One caution: even when available, the accuracy and reliability of "official" Chinese government lists may be of questionable value for trade magazine publishers. For example, such lists may contain high numbers of academic names (not separately identified) that might not represent a viable buying influence for advertisers.

Advertising

A recent report by the U.S.-China Business Council indicates that growth in advertising revenues for magazines in China now exceeds that of advertising revenue growth for television, newspapers, and radio.[8] The report does not say to what extent the revenue growth is derived from national or international advertisers. Nor does it distinguish between foreign or local publications other than to stress that the top revenue generators were major titles published in joint venture between Chinese entities and foreign publishers. Another report, by the Chinese Advertising Association placed the total magazine ad revenue in China at $860 million by year-end 1998, up 35 percent over 1997. The message is clear: Western-driven, quality print media is in hot demand by readers and advertisers across the Chinese marketing landscape.

For publishers of a business or trade magazine in China, the advertising potential will likely come from both the outside, with multinational marketers and from the inside, with Chinese companies seeking greater exposure and sales. The ratio varies depending on the nature of the magazine as well as how effectively the publisher is able to set up the advertising sales effort.

U.S. publishers who launch a Chinese edition would, logically, first target their established or prospective international accounts in the United States, Europe, or Asia-Pacific, positioning the magazine as an export medium. That's the "easy" part. The "hard" part is developing *ad sales support from the locals*. Chances are *both sides* of the equation are needed to be successful, according the hard-won experience of well-known foreign publishing leaders who have been in the China market for years. For example, Boston-based IDG launched *China Computerworld* in 1980 by relying mostly on ads from Western countries, but today counts about 90 percent of its weekly ad revenue from local companies eager to establish themselves as leaders.

Lack of sales tools and professionalism. One of the great frustrations faced by publishers in selling advertising for their Chinese edition, is the aggravating unavailability of useful data that defines markets and readership. Statistics provided by the official China Periodicals Association are not considered very reliable. Demographics are hard to obtain through government bureaus that protect the data like state secrets. Circulation figures have been largely un-audited, although, as mentioned, BPA counts a small but growing membership of magazines within the PRC.

Putting together a cohesive advertising sales presentation on a Chinese edition, that makes sense to sophisticated international marketers, is more difficult and time-consuming than foreign publishers are accustomed to. Even with a solid ad sales story, publishers may find it does not "play" on many calls to Chinese national or local advertising prospects. The concept of advertising value itself is relatively new to the Chinese business culture. Decisions to advertise or "support" a magazine may be based on factors wholly apart from ad effectiveness—like personal favoritism or politics. Though such factors are evident in the Western world as well, in China they are likely to be found more frequently. The picture is improving, however, as Chinese companies and ad agencies grow and expand their staffs with young marketing people and media buyers who have been better-educated in modern advertising and can respond professionally to an ad sales pitch.

One encouraging educational note for improving publishing professionalism among China's youth is a recent introduction of an MA program in publishing studies at Renmin University of China, in Beijing. The two-year curriculum was developed jointly by Renmin and Stirling University of the United Kingdom, working through the British Council in Beijing. The objective is to assist China's publishing industry in providing skills training for the transition to a market environment. The first term began September 1999.[9]

Selling advertising within China also requires legal compliance. Publishers permitted to accept advertising must have an advertising agent registered and licensed in China.

Competition

There are some 5,000 magazine agencies producing about 8,000 journals in China at present. Only around thirty magazines are done in cooperation with foreign publishers.[10] Several other foreign magazines' Chinese editions enter through the import route, mainly using a combination of controlled and paid circulation.

Most of China's internally produced publications have flooded into existence since 1985 when China's eye to a marketing economy began to open. These magazines serve a scattered assortment of audiences and purposes from consumer merchandising to communist party politics. Some still cannot or will not accept advertising. Those that do are generally considered weak by knowledgeable advertisers and ad agencies. Printing quality is poor. Circulation figures are dubious. Editorial content is sometimes based on puff pieces or the availability of described products for sale.

In spite of their flaws, some Chinese local magazines, aimed at particular industries or special interests, pose troublesome competition with their foreign counterparts when both are going after local advertising budgets. The local Chinese publisher invariably comes in with substantially lower ad rates, or offers huge discounts on the spot to get the business. Some advertisers can easily choose price over "quality."

Experience of Established Publishers

Since the 1980s a number of U.S. and other international publishing companies have launched magazines in the China market, targeting many segments, from agriculture to high-tech. Their hands-on experience is invaluable for those seeking to understand China's complexities and opportunities. Many of these publishers are members of the American Business Media (see ABM website: www.americanbusinessmedia.com.) Among them are:

Advanstar
Aviation Week & Space Technology
CCI Asia-Pacific Ltd.
Dana Chase
IDG
Intertec Publishing, A PRIMEDIA Company
McGraw-Hill
Meister Publishing Co.
Miller Freeman
Penton Media, Inc.
Simons-Boardman
Watt Publishing

Other Important Publishers Experienced in the China Market include:

Cahners
Emap
Hachette Fillippachi
IMAS
MediMedia Pacific, Ltd.
Hearst
National Geographic
Newswseek
Reed Elsevier
Ringier
Time Inc.

Time4Media
Twin Poplars Ltd.
Vogel Verlag
Walt Disney
Zenith Media Asia
Ziff-Davis

General Contacts and Resources

There are many excellent sources of information and contacts available for publishers interested in the China market. Among the most helpful are:

- *The United States-China Business Council.* This private, non-partisan group is the principal organization of U.S. companies engaged in trade and investment in the PRC. It was founded in 1975 and now serves more than 250 corporate members. The U.S.-China Business Council provides a multitude of services and activities focused on the business of doing business in China. It publishes the leading journal on China trade and investment, the bi-monthly *China Business Review.* The magazine is available to U.S. subscribers for $99 per year, and it also provides an online version. The U.S.-China Business Council also produces *China Market Intelligence*, a newsletter for members, and many other useful reports and guides with up-to-the-minute information on the China market.

  U.S.-China Business Council
  Suite 200
  1818 N Street NW
  Washington, DC 20036
  Tel: 202-429-0340
  Fax: 202-775-2476
  Website: www.uschina.org

- *U.S. & Foreign Commercial Service Offices in China.* As part of the Commercial Section of the U.S. Embassy in China, the FCS offers a wide range of services and assistance that facilitate marketing in China. These include counseling of business visitors, programs to identify agents and arrange business appointments, and help in locating information and getting started in China. The Commercial Section also supports trade shows and match-making events, American Trade Missions to China and Chinese Buying Missions to the United States. The FCS provides excellent, detailed reports and data, such as "Contact China," the comprehensive guide for doing business in the PRC.

U.S. & Foreign Commercial Service
U.S. Embassy
2 Xiu Shui Dong Jie
Jianguomenwai
Beijing, China 100600
Tels: (86-10) 6532-3831
(86-10) 6532-3431
Fax: (86-10) 6532-3297

Another key U.S. government contact in China is:

U.S. Information Technology Office (USITO)
C314, Lufthansa Center Office
50 Liangmaqiao Road
Chaoyang District
Beijing, China 100016
Tel: (86-10) 6465-1540-42
Fax: (86-10) 6465-1543

- *American Chamber of Commerce, PRC.* This Chamber represents some 600 American companies doing business in China. It is a primary point of contact and exchange for American business people in China, and serves as a bridge between U.S. business and Chinese government officials. The Chamber promotes the development of trade, commerce, and investment. It provides a host of programs and services, including speakers and briefings, business surveys, a monthly magazine, and detailed contact information on resources in the China market.

  American Chamber of Commerce
  Suite 1903, China Resources Building
  No. 8 Jianguomenbei Ave.
  Beijing, China 100005
  Tel: (86-10) 8519-1920
  Fax: (86-10) 8519-1910

Chinese Ministries and Agencies

- **Chinese Embassy in Washington, D.C.**
  2300 Connecticut Ave. NW
  Washington, D.C. 10008
  Tel: (202) 328-2500
  Fax: (202) 588-0032
  (China also has Embassy offices in New York, Chicago, Los Angeles, San Francisco and Houston)

- **State Press and Publications Administration**
  Department of Newspaper & Magazine Publishing
  Liu Bo, Director
  85 Dongsi Nandajie
  Beijing, China 100703
  Tel: (86-10) 6512-4433
  Fax: (86-10) 6512-7805

- **Publicity Department, Communist Party of China**
  Publication Bureau
  Deputy Directors: Song Zhenling, Wu Shulin, Zhang Xiaoying

- **State Council Information Office**
  22 Anyuanbeili, Chaoyang District
  Beijing, China 100029
  Tels: (86-10) 6491-2479; (86-10) 6491-2924; (86-10) 6491-2979

- **Ministry of Communications**
  11 Jianguomennei Dajie
  Beijing, China 100736
  Tel: (86-10) 6529-2215
  Fax: (86-10) 6529-2201

- **Ministry of Culture**
  83 Jia Dong'anmen Beije
  Beijing, China 100722
  Tels: (86-10) 6555-2035 / 6555-1934
  Fax: (86-10) 6555-1934 / 655-1959

- **Xinhua Publishing Services**
  Beijing, China
  Tel: (86-10) 6307-4590

- **State Intellectual Property Office**
  6 Xituchenglu, Jimenqiao
  Haidian District
  Beijing, China 100088
  Tels: (86-10) 6209-3289 / 6209-3276
  Fax: (86-10) 6201-9615

Useful Websites

American Chamber of Commerce: www.amcham-china.org.cn
U.S.-China Business Council: www.uschina.org
U.S. Consulate in Hong Kong: www.usa.gov/posts/hong-kong.html
U.S. Commercial Service: www.usatrade.gov
U.S. Department of Commerce: www.doc.gov
U.S. Embassy in China: www.usembassy-china.org.gn
U.S. EximBank: www.exim.gov
U.S. National Trade Data Bank: www.stat-usa.gov
U.S. Trade Representative: www.ustr.gov
Chinese Embassy in Washington DC: www.china-embassy.org
China Concept Consulting: www.chinaconcept.com
China Council for the Promotion of International Trade: www.ccpit.org
China Business World: www.chinadaily.net
China Big Yellow Pages: www.chiabig.com
China Books and Periodicals: www.chinabooks.com
China Vista; www.chinavista.com/business/home.html
China Yahoo: www.glochinese.yahoo.com
Cool China: www.coolchina.com
Complete Reference to China Websites: www.aweto.com/china
FortuneChina: www.fortunechina.com
Sinofile: www.sinofile.com
SinoSource Search Engine for Chinese Companies: www.sinsource.com
Surf China Engine: www.surfchina.com
Shaodaola Search Engine: www.zhaodaola.com

Notes

1. "When China Wakes," *Folio*, June 1998.
2. "Contact China," U.S. & Foreign Commercial Service, Beijing.
3. Anne Stevenson-Yang (co-founder Twin Poplars Ltd., Beijing), "Word Games," *The China Business Review* (May-June 1998).
4. Source: U.S.-China Business Council, June 2000.
5. "Contact China," U.S. Foreign & Commercial Service, Beijing.
6. Source: Much of the material regarding circulation is referenced from Anne Stevenson-Yang, "Word Games," The China Business Review (May-June 1998).
7. *China Market Intelligence,* U.S.-China Business Council.
8. "China's Publishing Industry," U.S.-China Business Council, June 2000.
9. Source: Announcement by The British Council, Beijing.
10. "China's Publishing Industry," U.S.-China Business Council, June 2000.

References

American Business Media, 675 Third Avenue, New York, NY 10017.
CCI Asia-Pacific Ltd., licensees of Time Inc.'s *Fortune* in China.
Daniels, John D., and Lee H. Radebaugh. *International Business Environments and Operations*, 9th ed. Englewood Cliffs, NJ: Prentice-Hall, Inc., 2000.
Hayes, Mike. "When China Wakes." *Folio* (June 1998).
Stevenson-Yang, Anne. "Word Games," *The China Business Review* (May-June 1998).
U.S. & Foreign Commercial Service. "Contact China" Report, American Embassy, Beijing, People's Republic of China.
U.S.-China Business Council. "China's Publishing Industry." June 2000.

4

Book Publishing in China

Ian McGowan

U.S. President Clinton's visit to the People's Republic of China in June 1998 recognized China's crucial political role in Asia, which was partly dependent on the sheer volume of its population, and partly on its substantial military technology. But the visit also acknowledged China's economic potential, both as threat and opportunity: it is a threat, because improving technology, communications, and changing government policy, together with a large and comparatively low-paid work force, make it a source of manufactured goods that undercuts Western products. It is also a threat especially in the area of intellectual property, where there have been piracies of Western books, software, and music compact disks. There is an opportunity because China offers the world's largest potential market, increasingly seeking access to Western information and technology, and—despite the average wages—with a developing consumer market, especially in the major cities. No one who has been stuck in traffic jams on Beijing's ring roads, seen the demand for personal computers and the money lavished on the "little emperors" created by the single-child policy, or wondered at the prices of the drab Western-style clothing offered in shopping malls can doubt that. China has recently been the world's fastest-growing economy, averaging 10 percent annually. Forty-nine percent of Beijing's secondary school students have a personal computer at home—usually a local product bought in the high-tech Haidian area (Edwards, Drury, and Liu, 1998). Despite the obvious tensions, China is changing economically and politically—at least in some areas—and it is too important for the rest of the world to ignore. This chapter discusses some recent developments in publishing, and their consequences both for China and for potential rivals and partners. It is a problem that little has been written in English about one of the world's major publishing industries, and that direct access to data is limited.

In general terms, the great cultural achievements and mechanisms of ancient China are known in the West. China invented the technology upon which, until our own time, publishing as generally understood, depended—papermaking in the second century, printing in the seventh, movable (ceramic) type four centuries before Gutenberg. We have a nodding acquaintance with Chinese achievements in calligraphy, illustration, block printing and so on, but few students of the book industry know much of such crucial mechanisms as the Press and Publication Administration, the New China Bookstores, Chinese copyright law. This chapter is an external view, based on teaching Chinese practitioners, and discussions of the issues with practitioners, academics, and administrators.

We may first consider the industry in terms of a conventional market. China has a population of 1.2 billion, perhaps three-quarters literate, some four-fifths living in rural areas, often far from the great urban centers that suck in the talent to their universities, research institutes, and official departments. The book market, although changing, has some of the characteristics of a developing country:

- a high proportion of information and educational publishing, as opposed to consumer books;
- limited competition in publishing and retailing, with, by Western standards, poor distribution and presentation of books;
- a demand for foreign-originated research and business knowledge alongside an inability to pay the originators' prices;
- a strong, at times idealistic, belief in the value of knowledge for personal and national development.

(All of these are heavily affected by the political aspects discussed below.) Following the disasters of the Cultural Revolution, in which schools and colleges were closed in 1966, it took a decade to restructure the education system and address the hiatus in the training of a whole generation. Along the way, English replaced Russian as the standard second language: there are many locally produced English magazines, books, and TV programs; Russian is now a rare occurrence. There are 14-15 million pupils in senior secondary or technical schools, with centrally set standards (British Council, 1996);

there are some 1,100 higher education institutions, with over 5.6 million students, many responsible to the State Education Commission, or central government ministries in specific fields. These are large numbers, but low participation rates by Western standards: in 1996, only 3 percent of the relevant age groups entered higher education (Edwards, Drury, and Liu, 1998), but the target for 2005 is 15 percent.

In 1979, there were 105 publishing houses in China, producing about 17,000 titles per year (Xu, 1994: 35). Now there are some 560, producing up to 154,500 in 2001, of which some 40 percent were reprints.[1] In 1994, according to UNESCO figures *(Book Facts 1997*, 1997), 15,100 came into the categories of "Arts" and "Literature," against 29,000 in the UK's broadly comparable total; roughly half of title output is in the social sciences. By value, the Chinese retail market at about $3-4 billion (Zhao, 1997) and is worth perhaps one-tenth of the U.S. market *(Book Facts 1997*). As a customer, China imported from the UK—one the world's great exporting nations—only £1.4 million in books in 1997 (Edwards, Drury, and Liu, 1998), making it fifty-fourth in the ranking of the UK's markets, well behind Thailand and Dubai, less than half of Russia or Taiwan. In 2001, it imported US$28m in books and US$32m in magazines. Although China is a large market, it has several notable features: the low number of publishing houses; the dramatic title rise (that still leaves output low in relation to population); with a total unit output of some 7.7 billion per year, this means average print runs in recent years of 55-60,000—astonishing to Westerners. (Despite celebrated runs of millions for individual titles, publishers say runs are falling as the market changes; the average in 1978 was 250,000.) Journal titles have doubled in twenty years to 7,600, a third loss making, with government subsidy for most academic and STM work. And despite the desire for foreign products and the attention of some Western publishers, it remains to a high degree a self-contained market.

What factors have shaped this market? And is "market" strictly the relevant word in this context? The extraordinarily turbulent history of China highlights several factors: its respect for learning and literate culture; its leisured mandarins and imperial examinations; its disdain for and disruption by the youthful cultures of the commercial and industrial West; its struggle with Japan (which, having embraced industrialism, was compelled after the Second World War to take a different route from China towards progress); the violent reaction against formal education in the Cultural Revolution (by 1967, title output had collapsed to below 3,000, compared to nearly 40,000 a decade before, and the export market in Japan and other Asian countries was lost (Chen, 1993); but especially the creation of the People's Republic in 1949. It was the Communist Party's views on the nature of the state, commerce, education, and the media that ultimately determined much of what we now see. (We might contrast the recent publishing histories of Japan, the disputed nationalist territory of Taiwan, Republic of China, or the recently reunited Special Administrative Region of Hong Kong. Taiwan represents a substantial but separate market, publishing 14,000 titles in 1997. Lacking United Nations recognition, it cannot join the Berne Convention, and it has no copyright agreements with People's Republic of China, but it has steadily revised its laws to broad compliance with Berne and TRIPS standards (*Publishers Weekly,* 1998). Written Chinese is largely in a standardized language, though with simplified characters in PRC (see below). Spoken dialects such as Beijing's Mandarin/Putonghua—common speech—and Hong Kong's Cantonese are mutually incomprehensible, partly because of the differences in tonal systems. At present, their media industries are largely separate.

Immediately after the founding of PRC, the Ministry of Propaganda created the huge Xin Hua (New China) Bookstore (Ze, 1995), which not only dominated retailing throughout China, with many thousands of branches, but ran the state printing plants and was the publishing organ of the Party and government (in practice, indistinguishable). Although the printing facilities were later hived off and the publishing entrusted to the People's Publishing House, the Party had established the basic structure and functions of publishing in the second half of the century. David Ze quotes (1995: 449) official pronouncements from various decades: (1958) "The purpose of publishing is to strengthen socialist education, promote proletarianism, and wipe out capitalism." In 1983, the Central Committee said, "Our publishing cause is fundamentally different from those of capitalist countries in that it is a part of the socialist cause under the leadership of the party and it must promote Marxism, Leninism and the thoughts of Mao Zedong." (Nowadays they would no doubt add, "to hold high the great banner of Deng Xiaoping Theory through the collective leadership with Jiang Zemin at the core." The formulaic nature of official pronouncements would do credit to a medieval supplicant to the Blessed Virgin.)

Publishing therefore did not exist for entertainment. Its function was, in the broad and narrow senses, political. It did not operate in a free market, but within a state-controlled socialist system. In the provinces, the subsidiary apparatus of publishing was established with clear lines of responsibility to their Publishing Bureaus, to sponsoring government ministries in whose subject areas (health, trade, railways) individual houses were to publish, or to universities and other official bodies. The historical origins and the preponderance of informational publishing accounts for the number of houses with, to Western eyes, surprisingly specialized roles. There are Presses named for the Coal Industry, Textiles, Water, the Ocean, and Electric Power. The People's Liberation Army (PLA) has some twenty publishing houses, mainly directed to military and technical matters, but including three medical houses and the PLA Literature and Art Publishing House. Presiding over the entire system is the National Press and Publication Administration (NPPA) responsible to the State Council for the supervision of all book, journal, and newspaper publishing and distribution.

Although its role is slightly changed, the NPPA, recently renamed the General Administration of Press and Publications (GAPP), is an immensely important body. While part of its function now is to facilitate relationships with Western publishers through its International Department, its primary function is to monitor and regulate, in effect to administer the political, propagandist system, if only negatively by preventing deviation. (Strictly speaking, there is no censorship.) It directly controls a dozen leading publishing houses that serve as models to the provinces. It regulates the Xin Hua bookstore system with some 10,000 outlets. It controls, in practice, many of China's plethora of publishing organizations, such as the Publishers Association, the Copyright Agency, the National Copyright Administration, with all of which it shares an address and, with most, phone and fax numbers. One important instrument of its control is the allocation of book numbers to state publishers. Whereas the Western view of an ISBN is of bibliographical and marketing convenience, a Chinese book number is an indication of legal publication, though in a sign of the times entrepreneurs have been buying numbers from officially authorized houses or in effect "packaging" for them, to circumvent the system for private profit. Selling ISBNs is now technically illegal (Ze, 1996).

The traditional structure under the NPPA prescribed specific publishing remits to individual presses, which are both a defensible allocation of scarce resources and a discouragement of content in unauthorized areas. Publishers, being subsidized in the provision of capital assets and operating funds, had no financial need to look beyond their boundaries; while employees, performing a social function and provided with housing and other benefits by the employer, had no immediate incentive to take initiatives. Much has changed since the mid-1980s through the impact of national economic reforms. While the organs remain state owned, the subsidies have been withdrawn in stages so that managers find themselves confronting many of the same business issues as Western publishers. The need to generate income has led to blurring of the edges of subject lists, as editors have seen opportunities to produce titles with market potential, to protect themselves against falling sales elsewhere. Since the mid-1980s, many editors have had a financial interest in the success of "their" titles, in the form of a bonus related to profit targets. Not surprisingly, this affects their views of press strategy and individual titles. While it would be too strong to say that the role of the commissioning editor as conventionally understood is fully developed in China, there is no doubt that in the more forward-looking organizations, groups of younger, well-educated, and enterprising editors are in practice redefining their role to embrace financial and marketing aspects of the process. Even before Mr. Deng's influential 1992 speeches in the Southern provinces, laying a new theoretical basis for reconstruction, the NPPA had asked publishing houses "to make economic reforms to achieve more profit, to meet the challenge from international book market" (Yu, 1997).

This causes strain within a house and its economic structure. Traditionally, books in China are cheap, not just in Western exchange, but in local terms, partly because of the principle of access to knowledge and partly because the structure did not have all the features of a commercial system. Ordinary paperbacks now cost little more than one U.S. dollar, and academic titles perhaps $5, though there are specialist exceptions and prices are not fixed. It is characteristic of both socialist and developing economies that the retail price of the product is a much lower multiple of the manufacturing cost than in Western nations: Chinese government guidelines (Ze, 1995) set out price categories, assuming up to 60 percent of retail price as the cost of raw materials and manufacture, some 30 percent as the retail and distribution margin, leaving about 10 percent to cover other costs and profit. This was not a market-led system, requiring large overheads to compete with commercial rivals for the favor of retail booksellers or consumers, or to generate maximum returns for owners. In consequence, the editorial role loomed much larger in the typical pecking order and staffing structure than the West has been used to for half a century. The editorial role, accounting for up to half of staff (Ze, 1995), was basically concerned with content—sifting, checking, refining, vetting. The author, of course, did not have a financial interest proportional to the book's sales. His fee was based on the number of copies printed and a manuscript fee based on the number of characters in the MS, providing an incentive to write at length. Royalties are now possible, though commercial authorship is hardly a profession. Overheads in this system were already partly taken care of by the state subsidy, and partly thought unnecessary, as various functions did not need to be paid for or did not exist.

Gordon Graham, former president of the UK Publishers Association, revisited Beijing in 1997 and recorded an illuminating exchange at the sixty-strong Publishing Science Institute, a research body (Graham, 1997): What could Chinese publishers learn from colleagues in Europe and America? "Management and marketing." And what could publishers in the West learn from their Chinese counterparts? "Social responsibility." This identification of their deficiencies echoes the lament of most commentators on publishing in Asia outside Japan, and at least those local publishers who have some awareness of Western practices. In a recent survey by Esther Pacheco (1997), director of a university press in Manila, scholarly publishers in twelve countries from Pakistan to New Zealand identified their most urgent problems in the areas of sales and marketing, promotions, and general management. As the economic reforms continue in China, more money and attention will be given to these.

The traditional channel of distribution was through the retail outlets of Xin Hua Shudian, with printings closely tied to their orders, and no returns; their wholesaling system had a tight grip on the order and physical distribution process, which was inefficient. Following the reforms, there is gradual introduction of information technology, the opening by some publishers of their own retail outlets and distribution systems, and the use of private bookstores, "the second channel." Competition for sales of educational books, and the clustering of academic publishers' retail outlets in Haidian Book

Street near the leading Peking [sic], Qinghua and Renmin universities is helping to sharpen marketing efforts. The import of foreign printed materials (worth over $100 million per year) is taxed and restricted to a few organizations, of which China National Publications Imports Export Corporation is the most important (Edwards, Drury, and Liu, 1998). Although Chinese books sell in the West in small quantities to niche markets, for content or as design novelties, China suffers like most of Asia and Africa because the West has little interest in importing either its knowledge or its imaginative writing. At present, China, partly hampered by the routing through official export corporations, sells only three million units overseas per year (Shen, 1998).

The question of "social responsibility" is a more difficult one. Publishers in developed countries are often impressed and embarrassed when they compare their motives and content output with the belief of developing countries in the power of information and ideas to change lives and shape nations. No one could dissent from that desire to do good. In China, major official and professional pronouncements continue to stress social value and high quality content rather than title output. A recent conference report by the deputy editor-in-chief of the People's Literature Publishing House emphasizes the "social benefits" arising from China's modernization drive and the stress on high values; but he goes on to highlight items publishers "should stand by no matter what":

> The principle of persisting in correctly guiding public opinion in pluralistic conflicts among divergent cultures and ideologies. Since commencement of reform and opening to the outside world, divergent ideological tendencies, cultural trends, value concepts and moral concepts have revealed themselves, and there have been pluralistic conflicts among divergent cultures and ideologies. China is a socialist country that must take the road of building socialism with the Chinese characteristics. The Chinese publishers are obliged to persist in correctly guiding the people forward with correct opinion. They must advocate healthy, lofty ideology and culture, in order to create a good environment of public opinion for the modernization drive, and give a moral support to the modernization drive. It is imperative to take precautions against the spread of negative influence of publication, in order to prevent unhealthy influence of tendency from obstructing the advance of the modernization drive." (Li, 1997)

This reminds us of the peculiar sensitivity of book publishing as a media industry. In China's current economic reforms, publishing is not in the vanguard of privatization, and there is some debate as to whether that exists at all in this sector (Ze, 1996; Graham, 1997). In opening up to Western investment, government agencies have greatly relaxed their restrictions on participation. A partner looking for a joint venture in general manufacturing will be taken very seriously. In publishing, rights deals for translations are usually straightforward (individual publishing houses can deal directly with foreign firms), copublication of individual works can be accommodated; but joint venture is treated with the greatest suspicion and, despite several gestures towards limited foreign involvement, there is no established example of full participation.[2] The reason, ultimately, is political: control of ownership and content matters in the media in a way that is irrelevant in light engineering.

How are these issues perceived in China? An extraordinary phenomenon, not directly accessible by most commentators, is the rise of the study of publishing within China. A recent survey reports that there are several dozen research institutes in publishing science, with about 500 research professionals, and that in the last two decades journals about publishing "have been springing up . . .like mushrooms after rain," now topping fifty. Many of these are concerned with theorizing developments in a distinctively Chinese context. David Ze (1995) has discussed the establishment of the editorial role in ancient China, and China's long use of block printing, in terms of their desire to establish "standard texts"—standard, in that if they were free of the errors of textual transmission that Shakespeare was subject to, they were also official selections of approved content. If Confucius was the first editor, he was also the first censor. Ze (1995) tells how in the eighteenth century the emperor himself, Qian Long, was chief editor of a collection, and how in ten years there were about fifty cases of book banning, with over 2,000 works destroyed. Western countries also have their histories of censorship, rooted in control of political and religious ideology rather than private behavior. But China's political and social history appears to have created the conditions for a distinctive relationship between the publishing media and society that has continued under control of the State Council and NPPA, and resulted in a cautious conservatism of content. Much commentary accepts these orthodoxies. The leading research body, the Publishing Science Institute, took as one of its primary tasks the development of a theory of "editology," to underpin the work of the industry "in the unique Chinese tradition" (Ze, 1995). This means serving the national interest, now identified as "socialism with Chinese characteristics."

Direction of content, then, is likely to be an issue for the foreseeable future; but an area in which there have already been obvious developments is intellectual property. We are told that there was some protection as far back as the eleventh century. There was a modest history of copyright legislation in the twentieth century, but this was overturned by the creation of the People's Republic, in which the private ownership, for profit, of works of the mind might seem anathema, although there were in fact some relevant administrative regulations. Given this political context, the lack of a legal framework, and the attraction of reproducing foreign works in the original or translation without the nuisance of negotiation or payment, there sprang up a series of sub-industries pirating Western properties, with special bookshop sections barred to foreigners. The International Intellectual Property Alliance estimated (*Bookseller,* 1996) that as late as 1995 U.S. losses alone were about $1.8 billion (of which more than half was in software, but books accounted for $125 million). A famous textbook example is Louis Alexander's *New Concept English,* published by Longman, but apparently selling in China in the late 1980s, in eight separate unauthorized editions, 10 million copies per year (Yu, 1997); since 1992, the World Publishing House in Shanghai has had exclusive rights, and has sold 5.6 million copies per year. Mr. Alexander has made celebrated appearances in China and collaborated with local experts on tailored versions of his work (Shen, 1998). The copyright climate already started to change throughout East and South-East Asia in the 1980s. A series of countries either revised or passed for the first time national legislation on copyright. This owed less to outbreaks of virtue than to increasing pressure from the Western property sources and their own sense of long term self-interest as they anticipated full participation in the international rights trade.

Chinese legislation moves slowly, but the first Copyright Law of the PRC, together with its Implementing Regulation, came into force on June 1, 1991. They are not long documents, and the official English translations from the State Copyright Bureau make fascinating reading, partly because so much is familiar: the law was plainly drafted with an eye on Western precedents and accepts the Anglo-American property model; speeches by senior administrators continue to stress its commercial significance (Shen, 1998). It rehearses familiar categories such as literature, photographic works, and computer software. It was ahead of American law in recognizing the author's "moral rights" of paternity and integrity. It is explicitly concerned with a property right (to exist for the usual "life plus 50 years"): "The copyright in a work shall belong to its author. The author of a work is the citizen who created the work" (Copyright Law, II, 1990: 2). It goes beyond most other national laws in specifying (III: 24) the "basic clauses" to be included in an author contract; and it specifies rather wonderfully (Implementing Regulation, V, 1990: 40) that where an author submits an unsolicited manuscript and receives no response from the publisher within six months, he is entitled to demand its return together with "economic compensation." (Though it is hard to identify an authenticated case of compensation changing hands.)

With laws so formulated, China was able to sign both major copyright conventions in 1992. Western rights owners saw loopholes in the right to translate or reproduce (though not publish) for teaching or scientific research purposes (Copyright Law, II: 22); and any law is only as good as its enforcement. By the mid-1990s, despite some gestures by the authorities, piracy was running at the levels quoted. It took the threat of U.S. sanctions against imported Chinese goods to produce a serious response. In February 1995, they signed an agreement on the protection of intellectual property, and U.S. threats were withdrawn in 1996 after further promises and evidence of pirate CD factories being closed. Other recent progress includes a 1994 White Paper on Intellectual Property, and a State Council decision on administration enforcement agencies; the Criminal Law was also supplemented. There is now an elaborate system of national and provincial copyright bureau, led by the National Copyright Agency of China (under the State Council), responsible for administration, registration, mediation, and prosecution in copyright affairs. Related bodies deal with professional associations, copyright research and promotion, and rights dealing.[3] Under the criminal process in effect since 1995, punishment can be imposed in proportion to the seriousness of the infringement, the extent of unlawful income, and the offender's record. Prison sentences may extend to seven years, plus a fine. The Supreme People's Court required major centers to set up intellectual property tribunals. In the first six years, they heard some 3,500 cases. Beijing Higher People's Court has set up specialized i.p. courts, offering expert assessment and fast-track procedures. In five years, the city has handled almost 1,400 cases, of which over 500 were copyright, 70 percent involving foreign parties (Edwards, Drury, and Liu, 1998; NCAC, 1998; *China Daily,* 1998). (In practice, the main area of print piracy is now in unauthorized reprints and translations of foreign journals, the object of official action in 2001.) There have been high-profile crackdowns. One man got life imprisonment in 1995 for pirating 700,000 copies of twenty books (*Bookseller*, 1995). There have been recent reports of further destructions of millions of pirated items (*Bookseller*, 1998b); and the 1998 Beijing International Book Fair had a display by the Copyright Administration of pirate works, and photographs of official destructions. It will plainly take time to get this under control, and to create in many formerly unregulated countries a habit of mind among publishers, librarians, and teachers that accepts the practical and legal implications of unauthorized use. Various Western agencies and experts have been working for a decade to create this climate. (An unauthorized publisher had the bright idea of getting a team of academics to write a book on the U.S. intellectual property dispute, called *China Can Say No!* In four months in 1996, trading on anti-American feeling, it sold 1.5 million copies [Yu, 1997].) Meanwhile, China has acceded to the World Trade Organization in 2001 and strengthened the copyright law.

Even before 1991, some of the most enterprising Chinese publishers were visiting the Frankfurt Book Fair to buy rights; some of those publishers on both sides who took the initiative in the absence of formal legislation have done well in the last few years, as they already had trusted business partners in place (and building trust is important in China). Legitimate Chinese publishers are themselves in competition with pirates, who hustle their editions onto the streets within weeks of authorized titles. There are celebrated examples of Chinese publishers paying serious money for rights for some business and fiction: Peking University Press paid a $50,000 advance for Bill Gates' *The Road Ahead*, which has sold 400,000 copies at $2.40; the Shanghai Translation Publishing House bought the rights to *Scarlett,* the *Gone with the Wind* sequel, and has sold over one million copies (*Publishers Weekly*, 1998). That translation was something of a test case in 1991-1992 because not only did they buy the rights from a Taiwanese publisher, but they got the NPPA to suppress seven unauthorized rival editions (Jiang, 1992). (The full-form Chinese characters superficially familiar to most of us, and used by Chinese communities in Taiwan, Hong Kong, and the West are unusual in PRC, which adopted simplified character forms decades ago. So in negotiating a Chinese rights contract, it is possible to subdivide not only by territory but also by character form.) The fact that the same rights purchases are regularly cited draws attention to a problem. Because of the legal framework, Chinese publishers of translations now have the added expense of negotiating and paying for rights in convertible currencies, which makes projects less attractive. One estimate (Xu, 1994) suggests that between 1993 and 1994 the proportion of science books that were translations fell from around 30 percent to 5 percent, and that there were even bigger drops in contemporary fiction. (Classic authors have the same royalty-free attraction to Chinese as to Western publishers.) Although individual Chinese publishing houses can negotiate directly with Westerners, much business is done through the public organization, the Copyright Agency of China, and other provincial agencies. More Chinese visit Western book fairs in search of rights, while the biennial Beijing International Book Fair, though still very much a Chinese event, has significant foreign participation (including, in 1998, Cuban, Tunisian, and Iranian displays, in addition to the more predictable foreign presences). Controversy arose during the 1998 Fair when the government banned the activities in PRC of two important Taiwan-based literary agencies, which represented many UK and U.S. companies: Big Apple Tuttle-Mori and Barden-Chinese Media. The official nominee replacement is the newly created Beijing Literary Agency, partly owned by the local copyright administration (*Bookseller*, 1998c), but the Western-style commercial agencies are increasingly active.

In the short term, the Chinese publishing industry has plenty with which to concern itself at home as it struggles to find its unsubsidized feet in a developing consumer economy that is gradually becoming more discriminating. One analyst suggests that distribution and related problems cause underperformance by 50 percent against reasonable expectation of market size, based on income and education levels (Shen, 1998). To the Western eye, much Chinese publishing is unimaginative, with poorly produced output, poorly distributed, under the leaden hand of official guidance for the foreseeable future. Yet some Chinese printers can produce books to the highest standards. Computerized typesetting of Chinese characters was developed in the 1980s (Jiang, 1997). Books remain relatively cheap. There are energetic and educated young staff in some of the most successful houses, willing to debate not, perhaps, the fundamental purpose of publishing, but at least the means to make it more effective. It says something about the nation's culture and priorities that just before President Clinton's visit the government celebrated the fiftieth anniversary of the Republic by investing 400 million Renmimbi (£30 million sterling/ $45 million U.S.) to create the largest bookshop in Asia.[4]

Despite the large home base, some publishers and policymakers are keen to develop the export of both books and rights, to earn foreign currency and raise China's foreign profile, especially through English-language output. They look to the non-PRC Chinese speakers of Taiwan and Hong Kong (with perhaps 27 million people) and the Chinese diaspora from Australia to Canada (perhaps another 30 million). Gradual relaxations and increased participation at major trade fairs such as Frankfurt and Beijing will cause business to increase. In January 1998, the conference of national press directors discussed a NPPA ten-year plan for the publishing industry to 2010, in which it is suggested that China establish five to ten major publishing groups and a similar number of distribution groups, as a way of restructuring the industry to meet changing conditions (Shen, 1998); the process is under way with, for example, the Chinese Publishing Group including nine national publishers plus distributors. Meanwhile, some of the most energetic houses are making strides in their own marketing and distribution. The first survey in PRC of Chinese reading and book-buying habits, conducted by the Chinese Academy of Sciences, recently found that liberalization and the increase in output had led to a decline in the reading of revolutionary literature and Chinese classics (and a lessened popularity of foreign literature after 1992); manuals and books on trade and economics are now the most popular (*Bookseller,* 1998d) with broadly educational publishing taking over 40 percent of the market value.

In imperial China as in the People's Republic, the philosophy and mechanisms of publishing differed greatly from those of the commercial West, with mingled qualities on both sides, as their respective industries responded to and helped shape their host societies. Chinese publishing counts for little at present on the world stage; but it would be a foolish person who would predict that in fifty years time its research output would not make it a major player—probably in English—in many sectors. And although the West may find its officially sanctioned editorial policies unglamorous, it would be wrong not to recognize that element that, often in practice as well as theory, state and developing-country publishing have in common: a belief in the power of knowledge and ideas as transmitted through books. Confucius says in his *Analects:* "The man of virtue understands what is moral, the small man understands what is profitable." The Chinese publishing industry may yet work its way to a happy compromise.

Notes

1. It is hard to judge between the various figures apparently authoritatively reported by Chinese official sources: from the Institute of Publishing Science, Shen (1997) refers to 110,000 titles in 1996; but PPA figures quoted by Shen (1998) offer 112,813. Some discrepancies seem caused by rounding, others by questions of definition.
2. Foreign publishers cannot operate directly as principals in China, and limited progress has been made on the implementation of five joint-venture licenses theoretically available since 1996 in the less problematic STM area (Edwards, Drury, Liu, 1998). Operating in the less political environment of Shanghai, Bertelsmann launched the first joint-venture book distribution, with local partners. Technically, it is a book club with 1.5 million claimed members, selling only books from official publishing houses, some of them foreign originals licensed in association with the local house. *Elle* is claimed as the first joint-venture magazine (*Publishers Weekly,* 1998).
3. A useful guide to this structure is published by the NCAC in English and Chinese: see Works Cited.
4. The Beijing Book Mansion in the Xidan central shopping area has 160,000 square feet of attractively displayed and widely varied stock (books, music, videos) with an Internet café and journals reading area. Personal observation recorded four floors rather than the eleven of the *Bookseller*'s report (1998a). A second superstore is planned for the tourist city of Xi'an.

References

Book Facts 1997: An Annual Compendium. London: Book Marketing Ltd., 1997.

Bookseller. "China and U.S. near the Brink" (February 17, 1995): 18.

Bookseller. "Seeing a Natural End to Piracy" (October 4, 1996): 18.

Bookseller. "China Opens First Book Superstore" (May 22, 1998a): 10.

Bookseller. "China Cracks Down on Pirates" (June 26, 1998b): 12.

Bookseller. "Chinese Ban Big Apple and Barden" (September 4, 1998c).

Bookseller. "Chinese Reforms Help to Boost Book Sales" (October 2, 1998d).

Bookseller. "Mixed Signals from Chinese Market" (June 18, 1999): 10.

Book Trade Year Book 1997. London: The Publishers Association, 1997.

British Council. *International Guide to Qualifications in Education,* 4th ed. Manchester: The British Council, 1996.

Chen, Yingming. "Chinese Publishers and the World Book Market." M. Phil. diss., University of Stirling, 1993.

China Daily (English language). "Court Handles Cases Faster" (August 31, 1998): 3.

Copyright Law of the People's Republic of China 1990, and Implementing Regulation. Official English translation by State Copyright Bureau, 1990.

Edwards, Chris, Russell Drury, and Liu Jie. *Publishing Market Survey: China.* 2d ed. Manchester: The British Council, 1998.

Graham, Gordon. "Beijing Interlude: Notes on the Chinese Book Scene." *LOGOS* 8(3) (1997): 140-5.

Jiang, Wandi. "Introducing *Scarlett* to China." *Beijing Review* (English language) 35 (30) (1992): 30-2.

Jiang, Wandi. "Liberating Chinese Printing with Keyboard." *Beijing Review* (English language) 40(19) (1997): 13-16.

Li, Wenbing. "Chinese Publisher's Social Responsibility Consciousness." Paper 5 in *What is Occurring Now in Publishing—Toward the 21st Century; Eighth International Forum on Publishing Studies.* Tokyo: Japan Society of Publishing Studies, 1997.

National Copyright Administration of China (NCAC). *The Copyright Protection in China* (Chinese and English). Beijing: NCAC, 1998.

Pacheco, Esther M. " Scholarly Publishing in Asia and Its Challenges." Paper 14 in *What is Occurring Now in Publishing—Toward the 21st Century; Eighth International Forum on Publishing Studies.* Tokyo: Japan Society of Publishing Studies, 1997.

"Special Supplement: China: The Once and Future Market." *Publishers Weekly* (July 27, 1998).

Shen, Chiang. "A Survey of Copyright Licensing between Britain and China in the Book Publishing Industry." M. Phil. diss., University of Stirling, Stirling, U.K. 1998.

Shen, Ju Fang. "The Evolution of Publishing in Modern China." Paper 9 in *What is Occurring Now in Publishing—Toward the 21st Century; Eighth International Forum on Publishing Studies.* Tokyo: Japan Society of Publishing Studies, 1997.

Xu, Mingqiang. "Book Market and Rights Licensing in China." In *The Emerging Markets of the Pacific Rim: VIII International Rights Directors Meeting*. Frankfurt: Frankfurt Book Fair, 1994, pp. 35–42.

Yu, Chunchi. "The Changes of Roles and Responsibilities of Editors in Publishing Houses in China. M. Phil. diss., University of Stirling, Stirling, U.K. 1997.

Ze, David Wei. "China." In *International Book Publishing: An Encyclopedia*, edited by Philip G. Altbach and Edith S. Hoshinopp. New York and London: Garland, 1995, pp. 447-460.

Ze, David. " The Future of State Publishing in China." *Bellagio Newsletter* 16 (1996): 15-17.

Zhao, Bin. "The Role of Hong Kong Publishing within the Chinese Language Publishing Industry: Its Opportunities and Limitations." Paper 7 in *What is Occurring Now in Publishing—Toward the 21st Century; Eighth International Forum on Publishing Studies.* Tokyo: Japan Society of Publishing Studies, 1997.

5

Scientific, Technical, Medical and Professional Publishing

Wang Jixiang

Historical Review of STM publishing in China

Four major inventions in ancient China are the symbols for humanity's transition from ignorance to forming a civilization. All four of these inventions are closely related with scientific technology and among these, the technologies of papermaking and printing established a firm base for book publishing in ancient China. From the earliest medical book *Huangdi Neijing* in the Zhan Guo Period to agronomist Shengzhi Si's book *Jisheng Zhishu* in the Xi Han Period, from Sixie Jia's book *Qimin Yaoshu* in the Bei Wei Period, and from the most important ancient mathematics book *Jiu Zhang Mathematics* in the first century B.C. to famous doctor Simao Sun's book *Qianjin Yaofang* in the Tang Dynasty, from Kuo Sheng's scientific technology encyclopaedia *Mengxi Bitan* in the Bei Song Period to the earliest botany dictionary *Quan Fang Bei Zu* in the Nan Song Period and Shizhen Li's medical book *Bemcap Gangmu* in the Ming Dynasty, all these books documented the improvements and development of science, technology, and medicine in ancient China. British scientist Joseph Lee stated, "The number of books published in China till the end of fifteenth century is more than the total number of books in the world." It is obvious that STM publishing in ancient China is unique and outstanding in the history of civilization's development.

In 1440, after German Johannes Gutenberg invented the printing press, the number of printed books increased geometrically both in

Europe and China. In the sixteenth century, many new printing or publishing houses were established and provided a favorable condition not only for religious but also for STM publishing. Many scientific and technological books such as Andreas Vesalius' *Structure of Human Body* and Copernicus' *De Revolutionibus* were published. In 1665, the earliest scientific and technological journal was launched in France. In the following years, scientific and professional journals represented about half of the share of STM publishing. This sudden availability of information pushed forward the development of science, technology, and medicine in many parts of the world. However, compared with the rapid development of science and technology in the Western world, China's development of science and technology and its STM publishing slowed down in the sixteenth century because of several historical reasons.

At the end of the sixteenth century and the beginning of seventeenth century, a new trend of translating, publishing, and collecting scientific and technological books emerged in China. The transition was reflected in two ways. One is that some STM books were abridged editions from Chinese scientists, such as *Nong Zheng Quan Shu, Tian Gong KaiWu, Ben Cao Gang Mu, He Fang Yi Lan, Wu Li Xiao Shi, Xu Xiao Ke Journal*. The other change was that some missionaries like Li Madou, Yuhan Deng and Chinese scholars like Guangqi Xu, Zhizao Li translated and published Western scientific and technology books, such as *Ji He Yuan Ben, Chong Zhen 0li Shu, Yuan Rong Jiao Yi, Tong Wen Zhi Suan*, etc. It is unfortunate that this trend of translating Western STM books ceased when, in 1723, Yong Zheng became emperor and expelled Western missionaries. China closed the door to the world at a time when Western technologies forged ahead rapidly at the beginning of the eighteenth century.

After the Qing Dynasty lost the fight with Britain, the Qing government was forced to sign the Hu Men Treaty in 1843. The five major cities of Guangzhou, Fuzhou, Xiamen, Ningbo, and Shanghai were opened as trade ports and Western missionaries entered China again. As they spread the Christian faith, they also took current printing technology and STM books into the country. The translation of these books made the knowledge and information of the Western world available to China. In 1860, after the Qing government signed the Beijing Treaty, China was forced to open itself even more widely to other parts of the world. Western missionaries established about sixty publishing houses in different cities in China. When they published religious books, they also translated many STM books. Facing the decline of the country, the Qing government began to pay more attention to Western scientific and technological knowledge and established many translation and publishing institutes. The Jiangnan Produce Bureau Translation House, which was founded in 1868, was the most famous of these institutes at that time. During the more than forty years of its operation, it translated more than 266 books, among which were 198 scientific and technology books, which occupied 74.4 percent of the total list.

When these early missionaries published books, they also published textbooks on religion, ethics, history, geography, and mathematics for their religious schools. The publishing of these books promoted the reform of education in the late Qing Dynasty. The *Meng Xue Textbook* was the first Chinese edited textbook by Nan Yang Gong Xue in 1901. Later, the Chinese government established Culture Press and Guangyi Press to publish a wide range of textbooks. With the establishment of the education reform, Commercial Press published *Newest Textbooks* in 1904 and became popular in all parts of China. The publishing of new textbooks not only changed the structure of publishing in China but also influenced the development of education in contemporary China. The translation and publishing of Western STM books and new textbooks promoted the birth of new subjects such as astronomy, chemistry, mathematics, mineralogy, physics, medical science and education in China.

After the establishment of the Republic of China, indigenous STM and professional publishing of books written by national authors developed further in China, but the main content continued to be translations of Western STM works. The National Government established a Central Research Institute, the first modern scientific and technological research institute in China at the end of 1927. Its research included a wide range of topics: astronomy, mathematics, engineering, meteorology, physics, chemistry, geology, geography, biology, agriculture, forestry, anthropology, archaeology, social sciences, and medical science. The Astronomy Research Institute, Meteorology Research Institute, Physics and Chemistry Research Institute, Geology Research Institute and Social Science Research Institute were established in June 1928. Later, these institutes published monographs and journals documenting their research and issued special monograph publications to report their achievements and to communicate their work with other research centers.

With the permission of the National Government, the State Peking Research Institute was founded in September 1929, under which eight research institutes (Botany, Chemistry, History, Pharmacy, Physics, Physiology, Radium and Zoology) were established. It merged with the State Northwest Agronomy Institute and the China Northwest Botany Research Institute. It administered its own publishing department, which was in charge of editing and publishing periodicals, special monographs, and collective works written in the institute. The department also reviewed publications from other parts of China and international research centers. Research reports and related documentation generated by this cluster of centers under the State Peking Research Institute were published in their journals and monographs.

The launch of the People's Republic of China in 1949 created a new, wide potential for the development of the country's economy and related fields of technology. After fifty years, research in astronautics, atomic energy, aviation, biotechnology joined the leaders of the world. At the same time, the study of general science was developed with a new emphasis. A new system of general science education was set up throughout the country and with it a large amount of general science reading was published. The quality of science was improved as a result of this new emphasis on the sciences in education. To confirm the reform and open policy, Xiaoping Deng proposed: "Modernization of science and technology is the key element for progress. If there is no quick development of the sciences and technology, there will be no quick development of the national economy." Under this national policy, the country spent about twenty years catching up with the rest of the world in the development of science and technology. During this historical period, STM and professional publishing experienced a great phase of growth and expansion.

Current Situation of STM Publishing in China

Today, there are 141 STM publishing houses or those that can publish STM books and journals. Most of these were established after 1982. Their publishing programs (topics, number of titles, revenue amounts, price, etc.) are documented in Table 5.1. We can note that the scope of STM publishing in China is still small based on the data of Table 5.1. Based on the total 141 publishing houses, only thirty-four publishing houses had a total annual revenue of more

Table 5.1
STM Publishing Statistics in China in 2001

Company	Kinds	Front Lists	Volume (10 thousand)	Page (thousand)	List Price (10 thousand RMB)
Science Press	2167	1208	3740	402865	50245
People Education Press	2263	505	5478	521443	49521
Higher Education Press	3253	1104	6309	881718	99215
China Agriculture Press	1399	809	1017	95891	17245
Popular Science Press	551	403	906	51196	8728
Machinery Industry Press	1828	1070	1007	171958	28764
China Construction Industry Press	854	336	328	83526	17045
China Railway Press	659	456	755	100097	15690
People Communication Press	672	325	377	65125	12327
People Post and Telecommunication Press	1094	791	675	146448	24276
Geology Press	374	263	1762	104672	13518
People Public Health Press	1053	501	1554	351447	46853
National Defence Industry Press	578	380	2069	68701	11181
China Water Resource Press	521	334	262	47796	8288
China Electrocity Press	614	327	324	65578	12207
China Finance and Economy Press	775	488	1281	179856	25501
China Light Industry Press	526	371	313	32682	8866
Coal Industry Press	151	125	136	12192	2719
Petroleum Industry Press	439	386	205	30775	5780
Atomic Energy Industry Press	68	66	29	2503	539
Science and Technology Document Press	349	227	276	35687	5161
Metalurgy Industry Press	291	233	119	16778	2901
Chemistry Industry Press	806	423	458	76142	14067
China Metrology Press	214	152	218	22348	4232
Ocean Press	262	252	186	24239	4678
Earthquake Press	124	116	75	11216	2157
Meteorology Press	268	237	189	14614	2842
China Map Press	561	145	9135	237316	40352
China Forestry Press	348	233	155	20014	4623
China Labor Safeguard Press	729	350	1120	110447	13862
China Goods and Materials Press	90	67	33	4260	707
Electronic Industry Press	1335	841	1349	266142	37560
China Textile Press	393	222	208	27225	6221
China Standards Press	283	258	120	30733	12770
China Medical Sceince and Technology Press	245	135	156	30353	5844
Priting Industry Press	36	17	34	3404	1024

China Environment Science Press	284	124	228	18034	2906
Intellectual Property Press	129	102	174	17013	3225
People Millitary Medical Press	404	256	196	35486	5695
Jindun Press	907	358	1672	116495	15611
China Agriculture Science and Technology Press	217	217	80	8526	1935
Astronavigation Press	33	29	38	3881	838
Weapon Industry Press	178	157	101	10706	1918
China Oil Chemistry Press	177	124	67	12120	2710
Aviation Industry Press	157	157	102	17383	2859
China Chinese Medicine Press	220	134	160	25922	3463
China Construction Material Industry Press	152	123	86	12401	2434
China Earth Press	73	67	69	10912	3205
Star Map Press	107	35	1699	33264	8485
Beijing Science and Technology Press	178	130	207	18119	3278
Tianjin Science and Technology Translation Company	278	195	130	11534	2652
Hebei Science and Technology Press	219	179	1437	64007	5891
Shanxi Science and Technology Press	214	124	123	10057	1719
Inner Mongolian Science and Technology Press	81	63	52	5376	698
Liaoning Science and Technology Press	407	192	188	24277	6865
Jilin Science and Technology Press	164	90	173	10669	1697
He Long Jiang Science and Technology Press	125	108	35	5294	2052
Shanghai Science and Technology Press	988	481	1875	184767	21798
Shanghai Dictionary Press	193	127	189	52566	12823
Shanghai Popular Science Press	261	138	239	22509	3917
Shanghai Science and Technology Education Press	878	331	4506	157683	18596
Shanghai Yuandong Press	354	198	314	29047	4490
Shanghai Science and Technology Education Press	357	192	279	33716	5138
Jiangsu Science and Technology Press	529	224	3082	131712	21743
Zhejiang Science and Technology Press	323	206	742	44370	6833
Anhui Science and Technology Press	370	214	432	41084	4822
Fujian Science and Technology Press	267	168	203	14568	2994

Jiangxi Science and Technology Press	275	143	843	45704	5335
Shandong Science and Technology Press	458	246	2154	110289	12499
Qingdao Press	489	188	2566	83562	9729
Henan Science and Technology Press	290	152	534	41397	4861
Central Plains Peasants Press	145	72	50	3723	517
Hubei Science and Technology Press	298	178	575	29215	3923
Hunan Science and Technology Press	489	304	491	60514	7660
Guangdong Science and Technology Press	462	267	308	24557	5393
Guangdong Higher Education Press	194	74	229	21302	2624
Guangxi Science and Technology Press	345	149	1516	62356	7848
Hainan Press	366	241	1248	85798	11644
Sichuan Science and Technology Press	226	156	263	25081	2641
Chongqing Press	601	334	9267	462503	41962
Chengdu Map Press	139	70	3060	7152	2866
Guizhou Science and Technology Press	110	77	327	14362	1633
Yunnan Science and Technology Press	153	137	87	8396	1790
Shanxi(3) Science and Technology Press	137	116	306	14853	2382
Qinghai People Press	148	138	507	32118	2949
Xinjiang Science and Technology Health Press	700	448	878	38978	4333
Jinan Press	215	103	701	43839	4504
Huanghe Water Resource Press	101	90	24	3629	1022
Shengyang Press	360	207	535	35596	6110
He Long Jiang Education Press	625	118	3028	106189	13446
Ningxia People Press	155	148	844	56107	5185
Xizang People Press	275	70	535	28927	3370
Shanghai Education Press	1701	546	4594	256980	26906
Peking University Press	1218	658	1080	152236	22397
Qinghua University Press	1270	692	1282	246431	33189
Beijing Normal Unversity Press	1391	254	4227	283516	31270
Beijing Aviation University Press	197	100	116	21426	3105
Beijing Industry University Press	169	114	121	14218	1797
Beijing Natural Science University Press	276	125	1065	57126	6280
China Xiehe Medical Unversity Press	100	80	142	32335	5377

Nanjing University Press	347	200	484	52709	6295
Zhongshan University Press	262	139	138	18076	2733
Zhejiang University Press	565	242	477	50866	6178
Fudan University Press	603	331	673	102771	12950
China Science and Technology University Press	213	159	117	16723	2051
Shanghai Communication University Press	497	285	513	54670	8537
Jilin University Press	99	98	61	8410	1014
Haerbin Industry University Press	195	81	86	12123	1477
Xi'an Communication University Press	298	135	352	41386	9033
Xi'an Electronic Science and Technology University Press	337	125	213	36208	4019
Shandong University Press	243	168	169	19956	2619
Tongji University Press	214	114	132	22126	3297
Huazhong Science and Technology University Press	548	323	464	61491	6579
Dalian Natural Science University Press	184	107	174	28729	4374
Northwest Industry University Press	174	113	144	17741	2089
Huazhong Normal University Press	506	300	1138	103171	10278
Huanan Natural Science University Press	354	208	287	36668	4675
Chongqing University Press	412	171	426	41834	5588
China Geology University Press	66	57	38	4114	628
Shanghai Medical University Press	23	23	10	1516	341
Huadong Natural Science University Press	224	121	153	23001	2662
Hehai University Press	161	127	134	13201	1741
Dalian Maritime Affairs University Press	91	49	28	4682	934
Northeast University Press	149	124	73	10332	1442
Haerbin Engineering University Press	88	78	53	5855	775
Haerbin Natural Science University Press	195	81	86	12123	1477
Northeast Forestry University Press	151	113	80	5659	717
Shanghai Chinese Medicine University Press	122	74	40	3894	769
Eletronic Science and Technology University Press	103	83	39	6363	829
Middle-west University Press	164	105	119	16726	2012
China Coal Unviersity Press	140	129	56	7717	1234

Southwest Communication University Press	129	94	46	5299	820
Southeast University Press	374	201	178	27319	4200
National Defence Science and Technology University Press	115	90	70	9023	1161
Shanghai University Press	129	111	132	13197	1732
Guangxi Normal University Press	374	198	2983	155891	13795
Tianjin University Press	314	181	180	29851	5080
Wuhan Natural Science University Press	187	102	129	19211	2180
Wuhan University Press	416	266	329	46162	5493
Beijing Medical University Press	182	94	218	33858	4875
	59423	31994	115961	9221952	1246029

than 100 million RMB in 2001, 24 percent of total publishing houses or sixteen publishing houses reported a total annual revenue of more than 200 million RMB, 11 percent of the total or nine publishing houses reported more than 300 million RMB; six publishing houses reported more than 400 million RMB. It is only the two publishing organizations of the Higher Education Press and Science Press whose output value is more than 500 million RMB. It is necessary to state that he Higher Education Press is not a STM publisher.

Table 5.2 presents the situation of STM publishing from 1998 to 2001. Based on this data, we can conclude that:

1. With the implementation of "use the education of science technology to develop the country" strategy, the amount of STM books published increased. In 1999, the total amount was 29,141 and increased to 34,126 in 2001, increasing 17 percent.

2. The list price has and continues to increase. In 1999, the average list price was 18.00 RMB, 20.32 RMB in 2000, and 21.15 RMB in 2001.

There is a shift in the different kinds of STM books published in that the number of natural science books increased and then decreased. The number of books released in the subject areas of mathematics, physics, and chemistry increased 33 percent from 1999 to 2001; books of astronomy and geoscience decreased 17 percent from 1999 to 2001; books of biotechnology increased 23 percent from 1999 to 2001; books of pharmacy increased 12 percent from 1999 to 2001; and books of agriculture were stable. The number of new titles in industry technology increased 23 percent from 1999 to 2001; books of communications and transport had little change; books of

Table 5.2
STM Publishing Subject Areas in China from 1999 to 2001

Category	1999			2000			2001		
	Titles	Print run (10,000)	Price (10,000)	Titles	Print run (10,000)	Price (10,000)	Titles	Print run (10,000)	Price (10,000)
General Concepts of Nature Science	741	4326	33712	926	3883	28727	893	3927	23091
Mathematics, Physics, Chemistry	2010	2380	30443	2530	2613	37202	2673	3138	45468
Astronomy & Earth Science	674	200	4430	630	188	4485	558	198	3968
Bio-Science	455	200	4918	514	299	6608	560	301	301
Medicine	5750	4663	99994	6329	4639	116095	6440	5282	134247
Agriculture	3328	2984	27856	3384	2821	32538	3281	2040	31058
Industrial Technology	14428	10512	257669	16267	11014	292757	17694	12354	336868
Communication & Transportation	1308	1239	22583	1161	1153	22145	1414	1192	25738
Aeronautics & Astronautics	84	26	804	90	38	1300	111	41	1403
Environmental Science	363	559	5298	400	251	4702	502	271	5682
Total	29141	27089	487707	32231	26899	546559	34126	28744	607824

aviation and astronautics increased 32 percent with the development of this area of technology; books of environment increased 38 percent representing the largest area of growth. A preliminary review of available data suggests that these trends continued in 2002.

Current Publishing of STM Journal in China

STM journal publishing in China developed rapidly when the open reform policy was implemented 1978. The growth was supported especially after Xiaoping Deng proposed: "Science technology is the productive force and it is the first productive force." The growth in journal publishing was supported by three major factors. First, the number of journals increased more than in any period in history. From 1981 to 1987, the average increasing rate was 11 percent. There were 2,951 different STM journal titles available in 1988 and the number increased to 4,420 titles by the end of 2001. New subjects and sub-subjects in the sciences were the main cause for the increase in the number of titles. Second, total circulation for these journals increased. The total printing amount was 401.38 million issues in 2001, 1.37 times that of 1988 and 1.76 times that of 1978. The third factor contributing to the growth was the improvement of content quality in the STM journals. The publishing of journals helped to change many research concepts, methods, and information availability and in so doing improved the productivity of science in China. Many of the STM publishing staff focused on exploring new subjects and pursuing new trends in science research. Some journals gained international respect for their high quality and unique subject areas. Sixty-four journals that were elected to join SCI in America. *Geology Journal* (English version) has become the most influential and most advanced journal in this field.

Content and manuscript editing standards of STM journals strengthened during this developmental period. Especially in recent years, Chinese publishers have tried to develop a higher level of editorial standards. More and more journals use international standards and editorial style manuals similar to those of the professional associations and societies in the Western world.

Electronic publishing of STM journals has become a recent development. The construction of databases is the most important part of this progress towards online publishing. Two examples are as *Title of Chinese STM Journal Database*, and *China STM Journal Document Database* (English Version DSI, A). These databases are the core of domestic and international communications of STM information. China STM Journal (CD-ROM) and Electronic Journal Press were established in 1998, publishing collective STM journals in CD-ROM format which became very popular in the market. More and more STM journals are exploring electronic versions, complementing the printing version, following the trends in Europe and the United States. All aspects of the transition to digital or online publishing shows that the revolution of publishing technology in China for STM journals is on the upswing.

The group of publishers of STM journals has become increasingly stronger over the past five years. For example, the Science Press group responsible for editing and publishing STM journals employs more than 3,000 people. There are 276 editing departments, publishing more than 270 journal titles every year. Generally speaking, with the prosperity and development of STM journals in the past twenty years, we can conclude that STM journals have been a necessary and important media for communicating science, technology, and medicine.

Introduction of the Three Most Competitive STM Publishing Houses

The research group of the "China Book Publishing Resource Database" evaluated and analyzed STM publishing houses (not including educational publishing houses) and based its statistics on the five-year time frame from 1996 to 2000. The result showed that the most competitive and major ten STM publishing houses are: Science Press, People Public Health Press, Publishing House of Electronic Industry (PHEI), People Post and Telecommunication Press, Machinery Industry Press, China Construction Industry Press, Jiangsu Science and Technology Press, Shanghai Science and Technology Press, Jindun Publishing House, and China Agriculture Press. The following text will provide brief introductions for the top three publishing houses.

Science Press

Science Press is the largest comprehensive scientific, technical, and professional publishing institute formed by the consolidation of the Pre-Edit and Translation Bureau of Chinese Academy of Science and Longman United Book Company, which was established in 1930s. Science Press was established in August 1954 and, in March

1993, the name of Longman Book Company was taken up again for active use. Science Press represents a long history and has developed strong resources over almost fifty years. Its main responsibility is to publish STM books and journals. Jixiang Wang is the president of Science Press Group.

In June 2000, the China Science Publishing Group was formally established including the original key company of Science Press. It is one of the six national publishing groups and the members include Beijing Hope Electronic Publishing House, Beijing Kehai Electronic Publishing House, China Science and Technology University Press, Beijing Zhongke Material Import and Export Company. The China Science Publishing Group is the only STM publishing group in China. Its goals are to identify, collect, publish, communicate, and distribute new science and technology knowledge/information for education, research, and related applications in industry. There is also a newly stated goal to spend ten years to turn the China Science Publishing Group into a high level, comprehensive, international science, technology, medical and educational media group and, therefore, a main part of China's culture.

Science Press has published 20,000 different book titles, and more than 300 journals (20,000 issues) in forty-eight years. The STM titles published by Science Press occupy12.5 percent of all science books published in China in the same categories and more than 180 journals, including twenty-nine English-language editions. In recent years, Science Press actively developed STM electronic products including video, software, CD-ROMs and online databases. Most of its journals are on the first level of quality in China and very influential in the world. Many of them are included by search systems such as SCI, CA, SA, and EI.

In 2002, Science Press published 3,037 books (including 1,519 new books and 1,518 reprints) and 184 journal titles as1,293 issues. The total annual sales volume was 377.48 million RMB. Science Press has nearly 600 employees. Branch editorial, sales and distribution offices are located in such big cities as Shanghai, Wuhan, Shenyang, Changchun, Chengdu, and Shenzhen. After the open policy reform, in order to strengthen international contacts and communications, Science Press New York Co. Ltd. was established in New York City and the Hong Kong Kehua Publishing Co. Ltd. in Hong Kong. These two subsidiaries have developed good cooperative relations with 196 publishing houses in foreign countries. The offices of National Science and Technology Language Examination and Approval Committee and Science Publishing Fund Committee of Chinese Academy of Science are also included in the activities of Science Press.

People's Public Health Press

People's Public Health Press is directly affiliated with the Ministry of Public Health and is the "central level" medical and health professional publishing house in China. It was established June 1, 1953, and today is the largest, strongest medical publishing house in the country. The Publicity Department and General Administration of Press and Publication of the Peoples Republic of China recognized and awarded this Public Health Press as one of the best publishing houses in 1993. In recent years, under the leadership of President Yiqing Liu and with the efforts of all employees, People's Public Health Press has developed from a large publishing organization into a quality imprint. It has established a high quality group of authors, a staff of excellent editors, a publishing management system, and an efficient distribution system. People's Public Health Press has established a multi-category, multi-level, multi-media, rational structure publishing organization that includes books, journals, digital audio and video products. It has become the most important medical publishing house in China specifically and in the Asia region in general.

Currently, the total sales volume of People's Public Health Press is 400 million RMB; therefore, it is one of the strongest houses in China. In an attempt to become one of the top-level international medical publishing houses it encourages Chinese doctors to write high quality medical books and provides medical students access to the best medical textbooks in the country. Today, its catalog includes the foremost medical books written by prominent authors with the most prestigious affiliations.

During the past fifty years, People's Public Health Press has published over 20,000 titles and a total of 670 million volumes. It publishes about 1,000 new book titles a year and distributes over 10 million copies. Its aim is to achieve an annual sales volume value of more than 500 million RMB. While its editorial goals are to publish the best medical textbooks, reference books and general medical readings about modern medical sciences and Chinese traditional medicine.

People's Public Health Press has a good tradition of publishing medical textbooks. In 1950s, it translated Russian medical textbooks for domestic requirements. Now it organizes the best authors to translate, write and edit textbooks from all parts of the world. Each generation of medical students grows up with the textbooks published by People's Public Health Press. In recent years, People's Public Health Press strengthened its management, improved the quality of its publications, developed its services and has become the only multilevel medical publishing house to satisfy a variety of readers' requirements. The brand of People's Public Health Press has become synonymous with high quality books.

Early in the 1980s, in order to adapt to education requirements and enhance the organization and management of textbook publishing, the Textbook Office of Ministry of Public Health was established and its office was set up in People's Public Health Press. In China it is the Ministry of Education that establishes the curriculum and plans for medical training for which the Ministry of Public Health then develops the textbook publishing program. It includes support from medical schools, professors' reviews, and editorial content development as part of the editorial development and publishing process. The new National Higher Medical Textbook Research Institute was established in April 2000 as a result of the development of a special educational system reform. The Ministry of Education, Ministry of Public Health and State Administration of Traditional Chinese Medicine and National Higher Medical Education Society participated in the reform process and in creating the National Higher Medical Textbook Research Institute. The prime responsibility of this new National Higher Medical Textbook Research Institute is to plan, organize, write, edit, and evaluate medical textbooks. Its office is located in the People's Public Health Press.

People's Public Health Press continues to be the prime center of publishing for all market segments and levels including medical degrees, master's degrees, college students, training school students, adults, professional health sciences students. Its content includes clinical medicine, anesthesiology, pharmacology, preventive medicine, stomatology, nursing, forensic medicine, and Chinese medicine. It has become a major publishing house in the fifty years since its establishment in 1953.

In 2002, People's Public Health Press provided 500 different textbook titles, among which were more than sixty titles geared toward the new twenty-first century Education Program and more than seventy titles that are the most important and significant medical textbooks in China. The print runs of *Internal Medicine, External Medicine, Gynaecology Science, Paediatrics Science, Anatomy and Embryology, Physiology* collectively are over 1.5 million copies.

After the Ministry of Education proposed the requirement to conduct bilingual education, People's Public Health Press imported many foreign textbooks and published low-cost paperback reprint editions so that medical students could acquire high quality English textbooks at reduced prices.

In 1995, People's Public Health Press started to develop audio and video products and electronic publishing. A new imprint identified as the People's Public Health Electronic Audio and Video Publishing House was established in March 2002. Titles such as *Medical Visual and Listening Textbooks of the Ministry of Public Health (VCD)* and *Medical CAI Textbook of Ministry of Public Health (CD-ROM)* are the main products of People's Public Health Press. At the end of 2002 there were 414 titles in VCD format and 116 titles in CD-ROM format available in the market. These products cover all aspects of medicine and are used in education, as clinical guides, self-study and research. Medical and nursing schools, hospitals, and the general public have provided excellent reviews for these multimedia titles.

People's Public Health Press establishes special cooperative programs with medical schools and large hospitals to produce electronic audio and video products. This method ensures that institutes and authors can collaborate to develop the subject or topic in new multimedia formats. Every product needs to pass many reviews from experts to guarantee good quality and learning verification.

The Ministry of Education, Ministry of Science and Technology, General Administration of Press and Publication, and Ministry of Public Health approved the People's Public Health Press' editorial programs and publications. More than twenty titles were awarded the Excellent Multi-Media CAI Achievement of theMinistry of Public Health, the Fourth Session National Excellent Education Audio and Video Products Award in 2002; the National University Excellent Textbooks Awards; the Seventh Session National Science and Technology Audio and Video Products Award; the Fourth Session National Education Technology TV Textbook Awards of the Chinese Medical Association. People's Public Health Press owns three journals: *China Medicine Journal, China Clinical Doctor,* and *Life and Health*, all of which are under the direction of the Ministry of Public Health.

In the future, People's Public Health Press will pay special attention to the depth of content and the range of subjects in its ongoing effort to improve its products. It also plans to publish more electronic audio and video products, while at the same time exploring and developing digital online information technology and the Internet in medical education to provide for the development of education in the future.

Publishing House of Electrical Industry (PHEI)

PHEI, established in 1982, is directly affiliated with the Ministry of Information Industry. It is also a central level professional publishing house, whose president is Zhigang Wang.

PHEI now has more than 200 employees, among whom 70 percent are professional technicians and 25 percent have professional academic titles. PHEI employs a high quality group of editors with doctorate, masters, and bachelor degrees. PHEI pursues the policy of "open the gate to do business," based on the high quality publications developed by their professional staff. PHEI makes a special effort to establish direct relations and cooperation with professors, experts and technicians from well-known universities, research institutes, and companies. Many of the authors and editorial consultants are scholars, professors, experts from the Chinese Academy of Science, distinguished universities, and Information Technology companies.

PHEI's strategy is to focus on three areas of publishing management: logistics management, information management, and capital flow management. Logistics management refers to making the editing, printing, packing, and distribution more efficient and thus cost effective. Information management alludes to efforts to generate, analyze and use quality information to strengthen the control and coordination of production and the operating process. Capital flow management refers to financial management to shorten the cycle of turnover, reduce inventories and make more efficient use of capital.

Since PHEI's establishment twenty years ago, it has published 5,500 book titles for a total 100 million units printed, and more than 100 titles have been granted special awards. It also published more than 100 audio and video products, and about 600 electronic or digital products. In the past few years, PHEI has published an average of about 600 new books and 300 reprints of backlist titles every year, and the total sale is 150 million RMB every year. PHEI published 230 CD-ROM titles, which sold 1.3 million units last year. In addition, in 2002, 200 books titles and ten electronic products for which translation rights were purchased from foreign publishers were published.

PHEI is the first publisher to step out of China and cooperate with U.S. Wanguo Science and Technology International Co, Ltd. and U.S. Fuguo Media Group to establish three information technology companies. Now PHEI has established a long-term cooperative relationship with more than forty foreign companies in Britain, France, German, Korea, and the United States. PHEI also has developed and maintains a mature website and its own Internet online bookstore. Readers can order its publications and products from the Internet.

The publishing programs of PHEI include computer science and technology, communication and Internet technology, basic electricity principles and related electronic technology. The formats are different levels of textbooks and training manuals, reference books, and electronic products for online information studies. They also provide research and development of software and online services.

Summary

The brief review of the three largest STM publishing houses in China provides a frame of reference in relation to the other 138 STM publishing houses that published 4,420 journal titles and 34,126 new book titles during 2002. This data confirms that China is really at the beginning of a whole new phase of development in relation to the needs of the country for current and timely quality information in the subject areas of science, technology, and medicine.

References

Institute of Scientific and Technical Information of China. "Chinese Scientific and Technical Papers Statistics and Analysis 1999." Annual Research Report, Beijing, 2000.

Science Press website: www.science.com.

Wang, Jixiang. "Science Press." In *A Century of Science Publishi*ng, edited by E. H. Fedricksson, London: IOS Press, 2001.

Zhang, Fenglian, and Li Li. "Improving the International Influence of Chinese Academic Journals." *Journal of Scholarly Publishing*, University of Toronto Press, (January 2003).

Zhang, Xingyong. "Impact and Opportunity Expected from Chinese Sci-Tech Journals after China Joins WTO." *Chinese Journal of Scientific and Technical Periodicals*, Vol. 12, 1 (2001): 6-9.

6

A Growing Children's Book Publishing Industry in China

Li Yuanjun

Introduction and Early Development

The children's literature publishing industry in China has been growing along with China's reform and economic development. When Deng Xiaoping started to promote the implementation of the reform and open policy in China in the late 1970s, the publishing industry responded actively. At that time there was a "famine" of children's literature in China, resulting from the severe shrinkage of the children's literature publishing industry during the preceding decades and leading to the lack of literature catering to 200 million children in China. In 1978, the China National Publication Administration convened a National Children's Literature Work meeting in Lushan, Jiangxi Province. During that meeting, the situation of the publishing industry in relation to the children's market segment was summarized in writing as follows: "...there were about 200 million children throughout the country with only 200 editors and 20 writers experienced and working in this sector of publishing." Only two specialized children's book publishing houses were operating in Beijing and Shanghai in China before 1978. The poor situation was definitely limiting the development of the children's book publishing industry and was not congruent to the spirit of Deng's reform and open policy. After that 1978 meeting in Lushan, specialized

The original text was translated by Zhang Zhuoran, member of the Master of Science in Publishing program, Center for Publishing, School of Continuing and Professional Studies, New York University, New York.

publishing houses were set up to focus on developing children's books in various provinces, autonomous regions, and municipalities directly under the Central Government. From the late 1970s to mid-1980s, more than twenty state-owned children literature publishing houses were opened and operated in the local regions or provinces. Thanks to various local governments' support for these publishing houses, each of these publishing houses was set up as a small or middle-size organization and has been developing fast since then. More books for children were published in wider formats and broader content, leading to the disappearance of the "famine" of children's literature. However, looking back in general, the children's books published then were not rich in content or of high quality in packaging and production and, in addition, they catered to readers only in certain areas.

The children's book publishing industry in China developed at an increasing rate as the country's economy grew in the decade from the mid-1980s to mid-1990s, in terms of both the types of books and the sales volume published during this phase of development. Two key factors contributed to this growth. In particular, the establishment of the nine-year compulsory education system in China and the gradual increase in the admission quota of colleges and universities year by year have provided great business opportunities for textbook publishers and related books for students attending primary and secondary schools. The growth in types and sales volume of textbooks and supplementary literature led to an increase in strength of children's book publishing houses in various provinces, autonomous regions, and municipalities. A select number of these publishing houses adjusted their business structure based on their economic strength and switched their focus to high-quality book programs including children's literature, popular science, general reference and inspirational books. Children's books published during this decade reached an unprecedented level in terms of their content, packaging, and printing quality. At the same time, as a result of the government's open policy, the Chinese publishing industry began to step outside the country, beyond Hong Kong and Taiwan, to exchange ideas and communicate with their counterparts in foreign countries, and participate in an increasing number of international book fairs around the world. Children's book publishers focused especially on the Bologna Book Fair in Italy and the Frankfurt Book Fair in Germany. During this decade, Beijing International

Book Fair, held every two years, also created the opportunity for information exchange and rights sales for children's book publishing professionals, authors, and illustrators.

In the mid- and late 1990s, as well as at the beginning of the twenty-first century, the Chinese economy moved more and more towards a market economy. The publishing industry gradually followed that trend to participate in this new emerging market economy, along with the reform of the Chinese state-owned enterprise system. During this transition, it became apparent that young readers were more self-aware, with a real interest in reading, therefore requiring more appropriate reading materials. Meanwhile, the state, adopting more policies for the publication of books, moved towards commercialization. Through publishers' efforts, many high-quality literature books were published in China's market. Several best sellers with millions of copies sold per new title in this period, including both domestic authored books, such as *The Frost on Long River* and *The Millennium Groan,* written by Chinese writer Yu Qiuyu, and imported and translated best sellers such as *Who Moved My Cheese?* and *Rich Dad, Poor Dad*. This improvement indicated a new era for the Chinese publishing industry. Furthermore, *Harvard Girl Yiting Liu*, which is about the growth of children, and the *Harry Potter* and *Goosebumps* series as imported and translated editions, have sold well especially as international imports. Few children's books written and illustrated by local authors have sold as well in China during this period. Children's book publishers in China still expect to pursue the best sellers based on their experiences with the *Harry Potter* series of titles.

Current Situation of Children's Book Publishing

According to statistics from the General Administration of Press and Publication of People's Republic of China (GAPP), based in Beijing, there were 562 publishing houses in 2001. Although there are only about thirty children's book publishing houses, more than 500 of the other houses publish children's books. The number of children's books published in 2001 was 7,254 titles, of which 4,433 were new front list titles for a total of 228,750,000 units, 757,450,000 printed pages, and 1,525,040,000 RMB revenues (list price); compared with that in 2000, this was a 3.57 percent increase in titles (3.67 percent in front list), 35.44 percent increase in units, 19.22 percent increase in printed pages, and 26.93 percent in the amount of list price. The retail sales volume increased 18.90 percent. Table 6.1 shows that the increase of many key factors were greater than China's GDP in 2001.

Three growth trends in the development of children's book publishing in China are:

- Large Potential Market

  Because of its huge population, China has the possibility of becoming the largest children's book-consuming country in the world. Once readers' purchasing interest in books is stimulated, the potential of the market is tremendous. There are nearly 0.3 billion children under fourteen years of age. Although most of them live in poor rural areas, continuing economic development, speedy urbanization, the One-Child policy, and traditional education concept combine to create an enormous potential for real development of this market. Therefore, focusing on this potential market and seeking a viable market share as the market grows is the basic goal for children's book publishers.

Table 6.1
China Children's Book Publishing Report (1998-2001)

Item	Measure	1998	1999	2000	2001
Titles	Number	6,293	6,111	7,004	7,254
Front list	Number	3,407	3,421	4,276	4,433
Print run	0.1billion	2.4	2.2	1.7	2.3
Print pages	0.1 billion	8	6.4	6.3	7.6
Total value (list price)	0.1 billion RMB	15.5	12.4	12	15.2

Source: Yearbook of People's Republic of China

Table 6.2
Children's Literature and Book Awards in China

Name	Jury	Frequency
Chinese Book Award	GAPP	2 x year
National Book Award	GAPP	2 x year
National Award for Outstanding Literature for Children	Chinese Writers Society	2 x year
Bing Xin Award for Children's Literature	Bing Xin Award Jury	1 x year
Song Quinling Award	Song Quinling Foundation	1 x year

- Improvement of Books' Quality and Market Concept

  Since the competition in children's book publishing has become more and more intense, publishers have to research the market, research both the parents' and children's psychology to improve the quality of books and establish the brands. At the same time, publishers realize that effective marketing and distribution can increase sales. As a result, publishers are in the process of strengthening their advertising, promotion, and marketing skills to attract more readers.

- More and More International Communication

  China's book publishing industry always emphasized its originality and nationality. After China entered the WTO, the market opened up and there are more and more international relationships developing worldwide. Many excellent foreign children's books were published in China by legal licensing. For example, *Harry Potter,* published by People Literature and Art Publishing House, sold more than 5 million copies in China. In 2002, Jieli Publishing House purchased the translation rights for the *Goosebumps* series, which is a bestseller in Western countries and sold 1.2 million copies in the first six months in China. Copyright licensing for translation and reprint rights responds to the interests of the readers, but it also points out the lack of originality in China, a country which it appears lacks a new generation of creative authors and illustrators. The current trend also opens the gate to foreign publishers, providing more opportunities for cooperation, co-publishing of translations. and new project development.

Sales

Sales of educational books including textbooks occupy the book publishing industry's the largest market share, which is about 60 percent; sales of children's books represent about 3.7 percent. Tables 6.3, 6.4, and 6.5 show that children's books are growing at a very solid rate.

Table 6.3
Sales of all Books in 2001 in China

Total Units (billion)	Changes in %	Total Sales (billion)	Change in %
15.651	0.73	92.093	8.61

Table 6.4
Sales of Children's Books in 2001

Total Units (billion)	Market Share %	Change in %	Total Sales billion	Market Share %	Change in %
0.581	3.7	14.71	3.43	3.7	21.97

Table 6.5
Sales of Educational Books in 2000

	Units (billion)	Market Share (%)	Change in %	Total Sales billion	Market Share %	Change in %
Textbooks for Colleges	0.425	2.7	1.57	53.49	5.8	12.66
Textbooks	8.405	53.8	1.57	381.24	41.4	4.06

In 2001, the *Yearbook of the People's Republic of China* reported that there were 0.29 billion children under the age of fourteen. In order to conform to global statistics, we extended this to include children under eighteen years of age. Thus, there are 0.36 billion children and young adults in China who are under eighteen. This is a huge number and to make it even more meaningful, it is a number greater than the total population of the United States. China sees the potential of this market based on that number. Of course, since a large percentage live in poor rural areas of the huge country, the possibility of realizing short-term income with related profits is not high, but the current 4 percent sales volume is considered too low.

The other major problem in the children's book publishing industry is that distribution channels are not effective and there are not enough ways to reach the growing market. Except for retail bookstores, most distribution channels are focused on schools and on direct family home sales due to the characteristics of the readers. It is quite common in China to sell children's books through the schools. However, China has established a policy of quality education, the "lighten the burden on children" policy, and relevant measures to reduce the sale of books in schools. Sales management has become a challenge. On the other hand, since community administration is not perfect, it is often not easy to realize the sale of books to the family. Direct sales are in trouble because of the variable quality of the books, irregular sales operations, and unfamiliarity with the different market segments. Online sale of books to parents via the Internet is still in the early stages of development.

The sale of children's books in China is very much influenced by the seasons, with the New Year Holiday and Spring Festival being the best selling times. Winter and summer vacations (February, July, and August) and Children's Day are also important times for sales of children's books. The fact that sales fluctuate in various months requires publishers and booksellers to develop effective marketing plans, and to organize their production schedules, warehousing, order fulfillment, distribution, and shipping.

Price is another factor that influences sales volume. The price of children's books ranges between 2 and 20 RMB and the biggest sales volume is for titles that have a list price between 6 and 8 RMB. Statistics show that 54.59 percent of children's books are priced less than 10 RMB and 84.91 percent are under 20 RMB. In general, the prices of children's books are fairly low within the country and extremely low when compared to the global arena. China cannot enter the next growth phase, including a more competitive and divided market, without a clear analysis of the children's books sector and, in so doing, develop related effective management and marketing strategies.

Enlarged Market of Textbooks and Reference Books

Textbooks and reference books are and will continue to be the main products in children's book publishing due to the increasing awareness of culture, effective communication, lightening the burden of basic education, and the extension of higher education. China's focus on education and its goal and policy of "use technology and education to develop China," addresses the needs of this current situation. Therefore, textbooks and references books will not decrease in number due to the "lighten the burden" policy, but will increase along with parents' hopes for their children to be the best in the future society. This attitude is reflected in the following:

1. Quality education books suddenly have emerged as a priority. Quality education has become a hot topic in China with the change from exam-based education to quality education and increasing requirements for high quality talents. The education institutions and publishing industry are exploring new ways to train people in order to respond to the change. As a result, relevant textbooks of quality for teachers and books on education are being published and their sales reflect this promising development. Some books such as *Harvard Girl Yiting Liu, Open the Gate of Imagination, Quality Education in the U.S., Escape from University, Fly U.S. and Do not Control Children* sell very well in China.

2. New textbooks are being developed in and for the market. In recent years, the policy, "use technology and education to develop China," has contributed to the development of China's education programs. In 2000, universities and colleges enrolled 2.21 million students, tech-

nical schools enrolled 31.02 million students and primary school enrolled 19.47 million students. In 2002, universities and colleges enrolled 2.90 million and will enroll 16 million in 2005. The number of middle school students, increasing in rapid numbers, will make all textbooks an ever-increasing big market. With the development of quality education, new textbooks are introduced in a wide range of topics such as *New Chinese, Information Technology, English* and *Music*.

3. With the exception of textbooks and reference books, children's books are divided into more than ten different categories. These categories include children's literature, children's popular science, children's cartoons or comics, baby reading, youth Reading, children' intelligence, children's arts, cards, crafts and handwork, and classical reading for children. The first four are the main segments, which occupy 60 percent of the market share. In recent years, the most influential books were published in these four prime categories. For instance, in 1999, the bestsellers in children's literature were the *Children's Encyclopaedia in China*, as well as children's popular science, and the *Harry Potter* and *Goosebumps* series. All the children's book publishers are trying to work within these categories to gain meaningful market shares and related income.

Other Categories of Books for Children

Development within the prime categories for children's books, including reference books, literature, popular science, cartoons and comics, and reading for babies, appears to have peaked with comparatively minimal sales growth. However, children's special education, supplemental texts, and skill cards have increased at better rates in units sold and revenues generated. In the past two years, this cluster of educational books increased 57.4 percent and 22.6 percent, respectively; and the sales of cards increased 35.41 percent and 20.15 percent, respectively. It is noticeable that the increase is based on creative new products that combine handcrafts with working and thinking, and are entertaining. They have a new format, with beautiful designs, are of different sizes, and often have three-dimensional pictures inside, which stimulate children's interest and imagination. Folding paper, stick picture mazes, and pocket books are examples of this category of publishing. Because children are active, curious, and good at mimicking, they find the new products more appealing than traditional books. A few examples of these new titles that predict a change in requirements and market demographics are *Calculation Cards* and *Babies' Intelligence Exam.*

Effects of Economy, Culture, and Education

It is obvious that China is a country with vast geographic space and a huge population. However, economic development, education levels, and regional cultures created by history vary in the different areas of the country. Book market segments and niches are also different. To keep abreast of the many and often subtle various markets and consumer requirements publishers need a good understanding of regional consumers and their unique characteristics. Normally, China is divided into six regions: Middle East, Middle North, North East, North West, Middle South and South West. Table 6.6 lists the differences in sales of children's books in each of these regions.

Observations: *Middle East* includes Shanghai, Jiangsu, Zhejiang, Anhui, Fujian, Jiangxi, and Shangdong; *Middle North* includes Beijing, Tianjin, Hebei, Shan(1)xi, Inner Mongolia; *North East* includes Liaoning, Jilin, and Helongjiang; *North West* includes Shan(3)xi, Gansu, Qinghai, Ningxia and Xinjiang; *Middle South* includes Henan, Hubei, Hunan, Guangdong, Guangxi, and Hainan; *South West* includes Chongqing, Sichuan, Guizhou, Nunnan, and Xizang. Please note that Taiwan, Hong Kong, and Macao are not included in the tables.

These differences are not independent from the realities of the marketplace and they are caused by regional GDP, population den-

Table 6.6
Children's Books Publishing in Six Regions

China Region	Titles	Percent of Total	Units thousands	Pages thousands	Value thousand RMB	Percent of Total
Middle East	2,679	43.4	83,800	286,076	509,060	44.6
Middle North	826	13.4	12,770	62,209	166,670	14.6
North East	578	9.4	10,800	45,528	126,230	11.0
North West	459	7.4	14,230	41,073	117,980	10.4
Middle South	1,183	19.2	29,060	86,843	154,000	13.5
South West	440	7.1	15,660	29,651	65,960	5.8
Total	6,165	100.0			1,139,900	100.0

Source: *2002 China Publication Information Material Collection*, published by China Labor Security Press, Beijing, July 2002.

sity, education, culture, and other key demographic factors. The GDP is the key factor as documented in Table 6.7.

Comparatively speaking, the Middle East is the most developed region in regard to economic aspects, business services, and education levels, all of which are high, and its book industry represents more than 40 percent of the total. It occupies the most important region of the country, including the province and city of Shanghai. Compared with other regions, the reading levels and the consumption level of books in this region are the strongest. Despite a better educated consumer, the growth of children's books may face difficulties because Chinese publishers have not been required to develop marketing and distribution capabilities. The potential for children's books would be greater if publishers would develop their marketing skills.

Beijing is the representative city of the Middle North region. Readers in Beijing, supported by their middle-income levels, have high education levels and related cultural development. Since Beijing is an ancient city with a great cultural history, sales of children's books increased rapidly with the development of the current information economy. However, in contrast, sales of books in the rest of this region were below the national level, despite the fact that quality and content requirements are comparatively high. Classics as "good" books, but not necessarily the popular type of books, are the main standard of consumers. Because of Beijing's special position as the city of national media, publishers are willing to choose Beijing as the first place to carry out their marketing and promotion programs, hoping to expand from there to a wider national campaign, which makes the Middle North an important region as well.

The consumer patterns of the Middle South region are between the Middle East and the Middle North, with pragmatic or "practical use" as the essential factor. Since some of the provinces in this region are not along the coast, some titles with traditional internal characteristics sell well here.

The economic development is slow in the North East region. This lack of economic strength limits the consumer purchasing power but not the related development of books. Sales of children's books are very strong in this region. The reason is quite simple. Although there is no money to buy entertainment and books, children are considered the people's future. Thus, although total sales in this region cannot compare with developed regions like the Middle East, con-

Table 6.7
GDP in Six Regions in 2000 in China

Region	Regional GDP billion	Change in %	Provinces	GDP Billion	Change in %
Middle East	3666.43	9.8	Shanghai	455.12	10.8
			Jiangsu	858.27	10.6
			Zhejiang	603.0	11
			Anhui	303.8	8.3
			Fujian	392.0	9.5
			Jiangxi	200.0	8
			Shandong	854.24	10.5
Middle North	1221.63	9.7	Beijing	246.05	11
			Tianjin	163.941	10.8
			Hebei	507.631	9.5
			Shan(1)xi	164.01	7.7
			Inner Mogolia	140.0	9.7
North East	974.33	8.7	Liaoning	466.83	8.9
			Jilin	182.0	9.2
			Helongjiang	325.5	8.2
North West	453.712	8.9	Shan(3)xi	166.1	9
			Gansu	98.3	8.7
			Qinghai	26.312	9
			Ningxia	26.5	9.8
			Xinjiang	136.5	8.2
Middle South	2515.427	9	Henan	512.6	9.4
			Hubei	427.632	9.3
			Hunan	369.188	9
			Guangdong	950.604	10.5
			Guangxi	203.555	7.2
			Hainan	51.848	8.8
South West	870.632	8.5	Chongqing	159.0	8.5
			Sichuan	401.03	9
			Guizhou	99.332	8.7
			Yunnan	199.528	7.1
			Xizang	11.742	9.4

Source: *Yearbook of People's Republic of China*, Beijing, 2001.

sumer interest and enthusiasm are much higher than in other regions.

The North West and South West are the least economically developed regions. Their book publishing industry is not mature. The market is not stable. Sudden increases and decreases occur as these regions go through the variables of development. The reason is that in this area education levels are low, the density of population is spread out, and thus children's book publishers are weak. The regions are also far from the main distributor's network so that there are more steps involved to bring books from other regions. At present, major children's book distributors are not interested in these regions. Concentration on this market will require large investments and risk, but ignoring this region will create future problems for the market share and related growth potential.

The ratio between children in rural and urban areas has a great effect on the sale of children's books. More than 60 percent of rural children can only afford textbooks. Most of the other books for reading are rationed out by the government or donated by sectors of society.

Present Problems and Challenges

Children's book publishing should be the main force in China's book industry. However, children's book publishing occupies a fairly small market share because of the following problems.

1. The use of repeated topics and a lack of creativity in children's books are serious problems. There are so many same topics or similar books published. On one hand, publishers provide multiple choices to readers, but on the other hand, they force readers to choose from many similar books that contain the same subject matter. When readers choose a book, they are not likely to buy others on the same topic or with a similar story. Though publishing houses publish many books, sales do not increase because of this lack of uniqueness and creative new subject areas. According to a survey done by Beijing Kaijuan Book Market Research Institute, there are 349 versions of *Anderson's Fairy Tales* and more than 500 versions of *100 Thousand Why* available from various publishing houses.

2. The lack of original creations and failure to develop front list titles are the basic reasons for the slow-increase in sales. Although purchasing translation rights licenses or co-publishing titles may conceal this problem at a certain level, the problem will not be solved until the

issue of the development of new lists, new books with new ideas is addressed.

3. To generate sales volume, more and more publishers are using discounts, special pricing, and other low price incentives. The competition of low or reduced prices not only damages publishers' financial strength but also creates problems for the distribution channels. It also causes difficulties for investing in the future development of new and different publishing programs.

4. Lack of effective marketing, including all the elements used in the Western world is another problem for children's book publishing in China. Since consumers face multiple choices in books and other consumer products, effective marketing becomes essential to develop sales. However, China's book publishing industry entered the marketing economy only recently. At this stage of its development there is insufficient management experience, human resources, and a naïve working understanding of marketing with all its related practices. There is an understanding of consumer marketing principles, but not how these apply to advertising, promotion, publicity, direct mail, catalogs, sales representatives and online websites relevant and applicable to the publishing industry.

5. Market research is not well understood and thus not often used, which hinders the development, publishing, and supply of children's books in China. The supply of books is greater than the demand in the prime six market segments as outlined in Tables 6.6 and 6.7 Consumers' interests, requirements, and purchasing power need to be defined, based on careful and professional market research.

6. The establishment of brands has lagged behind when compared to what is seen at such international events as the Bologna Book Fair or the Frankfurt Book Fair. It is understood that brands directly influence consumer buying, but again, Chinese book publishers have not yet honed the skills and capabilities needed for brand development and maintenance, especially if they hope to purchase the rights to important titles that have world recognized brand identities. When consumers choose books, marketing is important, but a publishers' brand also helps consumers with their decision and ultimately increases sales. Unfortunately, at present, most publishers still focus only on marketing in general, which can hinder further development of children's book publishing.

These problems that the children's book publishing industry faces are the same as those that other sectors of book publishing will encounter. They are representative challenges that the entire publishing industry in China has been experiencing and the challenges will increase as the industry goes through major transitions during its next five-year phase of development.

The Future of Children's Book Publishing in China

Despite the number of problems that the market for children's book faces, it is apparent that the future market for children's books will continue to grow because the children's book market is the largest in the world. Most of the problems mentioned will surely be solved because of the huge number of the potential consumers, the increasing purchasing power of the people, the special attention of parents to children's education, international communications, and the openness of China's publishing industry. It has been predicted by the experts in the publishing industry that the sale of children's book will increase 100 percent in ten years, a fact which will provide a viable market and future for those publishers who want to enter this field. It is relevant to conclude this chapter by highlighting some key factors that support this optimistic view of future growth for this sector of the publishing industry:

1. The whole book consuming level is still comparatively low. According to recent statistics, China's book sales will maintain a minimum growth rate of 10 percent each year over the next five to ten years, which means that total sales will double in ten years. Concerning children's books, the estimate that sales will double is a conservative prediction.

2. Publishers' awareness of the market segment grows stronger each year. The subsidy from the government will be reduced and, therefore, publishers will have to depend on the market to survive and develop improved lists as well as their added value skills, including marketing and distribution. Under these circumstances, "watching the market" will no longer be a slogan; action will be required to establish a viable role in the projected growth of children's book sales.

3. Publishers are beginning to establish brands. In recent years, only a few publishing houses invested heavily to establish brands and they have now taken advantage of their recognized presence in the market. Although the children's book industry in general was gloomy at the close of 2002, these publishers with effective brands and related marketing management were still able to achieve double digit growth in sales. We believe, with the maturity of brand strategies not only for specific publishing houses but also for book series and special characters, there will be more opportunities for development and choices in the market.

4. The support of and benefits from professional market research bring an important and obvious added value to managing the publishing process. The Beijing Kaijuan Book Market Research Institute, which was established in 1998, is one of the first professional market research institutes in the publishing industry. Realizing the importance of market research, publishers of children's books have begun not only to adjust internal information systems, but also to draw support from external market research on consumers and niche markets. It is hoped that these efforts will restructure product/title development and focus marketing resources on the whole children's book publishing industry. Therefore, publishers will address many of the past problems across the country to develop this important sector and accommodate China's children so that they may read at home as well as in the schools.

5. Publishers' interests in gaining the necessary skills of marketing are increasing. More and more publishers realize the effects and value of marketing not as a general concept but as the complex integration of all related elements from advertising, promotion, and publicity through to sales representation. Leaders in the industry have begun to gather experience in marketing methods and they have achieved a measure of success. These are the leaders and now at their heels come the followers who wish to enjoy the same benefits of profitable growth while pursuing an increased market share.

Children's book publishing industry in China is developing at a rapid pace and its tremendous potential market is expected to become an important part of the global children's book publishing industry. The world will see a huge market for children's books emerge in Asia within the next five years.

7

A Study of Chinese Young Adult Reading and Its Market

Lin Chenglin

Introduction

What data exist in reading surveys, and their major conclusions, do not totally agree with pure commercial market research. Most of the reading surveys provide outlines of information or trends in reading; their results cannot easily be converted to be helpful for commercial decisions. However, reading surveys are the usual practice for the book industry to develop basic market information and it is no exception in China. Among various reading surveys carried out in the past year, those related to young adults in China have been quite popular as compared with those of other age groups of readers.

My study of Chinese young adult reading is based on a number of related surveys, including those done by China's book publishing trade publications, China Book Business Report (CBBR), as well as major surveys from other Chinese educational resources.[1] The following study is intended to identify young adult reading habits along with an analysis to provide a frame of reference for what is unique in China. And based on a small special survey by CBBR, this study will also try to clarify the market size of young adult reading in China.

The Size of the Chinese Young Adult Reading Sector

China as a country has the largest population and, therefore, it also has the greatest potential market for young adult readers. This is what people generally believe; however, in terms of specific numbers, it is really difficult to determine accurately especially for a group of readers made up of young adults. For one reason, one has to calculate population segments from the ever- changing number of the total population. Another problem is that the "young adult" is a somewhat strange category in dividing up the whole population or description of demographics. The Chinese government and book industry are not used to a concept called "young adult." For the purpose of this report, we have established an understanding for "young adult" to be that group whose reading is neither dependent as in primary schools nor independent as most of them are not living by themselves. The age range of the group is 13 to 19, and it might be older or younger by one year to be in this range. In trying to compare it with what we are used to, I find the Chinese students in "middle school" grades to best suit the young adult reading group category. The student starts his or her school at the age of 7 or 8. After six years of primary school he or she usually goes for three more years of compulsory education and another three years of either high school or professional school before going to college and university or entering the labor market by getting a job.

Let us first calculate from the most recent Chinese census reports for the size of the population segment that includes those aged 13 to 19. The Fifth National Census of China released its preliminary results on March 28, 2001. According to the official reports and statistics, mainland China has a population of 1.265 billion (1,265,830,000). The group aged 0 to 14 has 289,790,000 in number or 22.89 percent of the total population. The group aged 15 to 64 includes a total of 887,930,000 in number or 36.09 percent of the total population, and the group aged above 65 is 88,110,000 in number and 6.96 percent of the total population. Since the last national census took place in July 1990, the population in China has increased by 132,150,000, with an annual growth rate of 1.07 percent. The urban population of 455,940,000 takes up 36.09 percent of the total population.[2]

It is safe to say that the population grew at a moderate rate, and at a lower rate compared with other historical periods. Also, during the last ten or twenty years, there were no large social events that could cause large fluctuations in China's birth rate and thus the number of the newborn every year. Therefore, it is appropriate that we then could divide the 0 to 14 group by 14 in order to identify the number of Chinese people who are of the same age in the group aged 0 to 14. The result is a population segment of 20,699,286, or about 20 million. As the distribution of the young of various ages is relatively even, we could calculate the number of Chinese people aged 13 to 19 to be about 144,895,000.

Not all of the 144,895,000 young adults are valid in calculating the number of people in the "reading" group aged 13 to 19 because some of them are not even able to read. We must exclude most of the young adults who are not valid in our calculations because they live in the rural areas. Almost all the publishing industry surveys do not include readers in rural areas based on their poor consumption of books and other publications. In order to calculate the meaningful population segment, we have to focus on the urban Chinese population, which take up 36.09 percent of the total population. Based on such further adjustment, the number of members in the young adult reading group is an estimated and revised total of 52,292,606.

Although we can identify the number of young adult consumers, it is still not possible to use the 52.3 million for assessing the current market size. A number of unique factors need to be taken into consideration. For one thing, all of the reading surveys that include young adults are based on urban middle school attendance data; we have their amount of book consumption. But the middle school students cannot fully represent all the Chinese aged 13 to 19. When urban students finish their compulsory education, which ends at the student's age of 16, one-third of them continue middle school education in order to have a chance for attending a college and university. The other two-thirds either go to a professional school or join the work force. The Chinese book industry traditionally ignores the buying power of those who do not go to middle school, including grades 4 to 6, because those students who could not continue education or choose professional schools for whatever reason are too poor and too busy to read. Those who read after class among the group include about 34 million in number (67 percent of all of the urban young adults). In CBBR's recent survey of 2,219 middle school students from Shijiazhuang,[3] a medium sized and medium income level city of China, we have found that 55 percent of them spend 1 to 15 RMB yuan every month, 26 percent spend 16 to 30 yuan, 6 percent spend 31 to 60 yuan, 1 percent spend 61 to 90 yuan, and the balance of 12 percent of students never buy a book for out of class reading.[4] The survey found that the young adult reading segment of the population spends only an average annual amount of 245 RMB yuan on books, periodicals, and newspapers.[5] Such an amount could buy ten books in China, according to the average book price. We come back to the challenging question of how large is the Chinese young adult reading market? We at CBBR believe that the segment of the young adult reading market is around 8.33 billion RMB yuan or a little bit more than 1 billion U.S. dollars.

Learning and Reading: A Battle for Time

Young people in the age range of 13 to 19 should have established their own tastes for reading, and we need to determine what has been the biggest obstacle in deciding or developing reading taste. Most of the reading surveys that are based on young adults are from an educational perspective and actually obtain samples from schools with assistance from teachers. The phenomenon proves that in China young people's reading is mainly driven by education. It is true, on one hand, that the education system influences young adults or middle school students a lot. However, on the other hand, we could gather relatively limited information on Chinese young adults' reading in relation to reading for entertainment or for pleasure. Young adults have little power in selecting and purchasing reading materials on their own. In the traditional culture, the way to get educated is to read many books; education in China still means book reading (DU SHU'). However, it seems that education has become the number one obstacle in expanding Chinese young adults' reading. Therefore, what I will do in this chapter is to examine the pressure from education that seriously affects and decides the reading of young adults in China.

In a market segment of over 1 billion U.S. dollars, it is, of course, not feasible to limit readings only to education, but actually a large portion of the money middle school students have is spent on what we call the supplementary reading for textbooks. In our survey, around 20 percent of the students read only the supplementary books after school. Supplementary titles are all intended to prepare students for exams. In the past ten years, this unique type of book publishing had become extremely profitable until last year when the Ministry of Education ruled that the schools would not be allowed to recommend supplementary reading materials to students. The policy is called "reducing the burden on students" and intends to free students from hours and hours of studying after school. The policy has not been totally effective in freeing students from their overload of studies due to many factors that are deeply rooted in a system that has existed over the past twenty years.

Towards the end of 1970s, the primary school and middle school education restored its normal teaching after ten years of chaos, which is also known as the "Great Cultural Revolution." In 1978, Chinese colleges and universities began to enroll students, large numbers of youth of different ages. These young people began to try to change their fate through receiving a formal higher education. College students had been such an honored and highly respected group that the rejection ratio can be as high as one qualified student out of seven or eight candidates. To have their children go to college and university has been a dream of most of the Chinese families. After 1989, college students lost most of the favors from the government such as a system of assigning decent jobs for graduates, free tuition, and offering easy permanent urban residency for students from rural areas. Despite the removal of these benefits, the desire of going to college remained the same. To graduate from college was in many cases also the very first step, although not always the decisive one, toward success through studying abroad and postgraduate studies. If a middle school student failed to pass the annual entrance examination for college, he or she would lose most of the hope for a career. At that time, the enrollment capacity for college was one out of three or four. In the last five years or so, Chinese college students had to pay a considerable amount of tuition that would cost a family an equivalent of ten years of savings. Colleges and universities began to enlarge their enrollment numbers to around 2 million a year from 0.5 million in the early 1990s. Now half of Chinese students who finish twelve years of primary and secondary education will go to college, which is certainly enough for maintaining the level of competition. Despite the expectation to obtain better jobs and the need to pay high tuition, Chinese families are still enthusiastic for higher education and the competition is further intensified. The problem is that they know that without a college diploma, the young person will face even more competition in looking for a job in the lower sectors of the economy.

It is safe to say that Chinese young adult readers may be the busiest and hardest working students in the world. Many surveys of students' studying time have shown that they have very little leisure time they may spend on after class reading. In CBBR's reading survey of middle school students, other than school hours and resting/sleeping hours, 28 percent of students have 31 to 50 hours per week for them to control, 45 percent have 21 to 30 hours, 20 percent have less than 20 hours, only 7 percent have 41 to 80 hours. During the limited hours, one-third will be spent on doing homework, and if students are in the last year before graduation, this portion of time will be as high as 80 percent to 90 percent of their time studying. How much time is left for reading? The survey shows that 24 percent of students have less than 2 hours for reading, 56 percent spend no more than 8 hours for reading, and only 9 percent enjoy reading for more than 15 hours. And the survey also shows that 41 percent of school teachers and 46 percent of parents have a negative attitude toward increasing leisure reading, their main reasons are for fear of decreasing the student's ability to cope with the competition in taking exams and entering university education. It is apparently true for Chinese young adults to read more they need more time outside of their perceived and real pressures to pursue their education. A governmental survey has shown that 24.5 percent of Chinese students never read books other than textbooks and their related course supplements. However, there are other obstacles.

Chinese families, like in many countries in Asia, have the tradition for emphasizing education and are usually no less powerful than schools on the matter of deciding how students should spend their leisure time. They even control much of the students buying power for books, periodicals, and newspapers. In our survey, 55 percent of students spent less than 15 RMB yuan a month for books, which could not buy even one copy of an average priced paperback. The books, as well as periodicals, are more and more expensive each year.

Another major obstacle (other than the economic, educational, and family) for students to get more opportunities to read are the poor school libraries and public libraries. The libraries that students could visit have been visited less and less, only 1 percent of the Beijing students routinely go to a library. Beijing has 3 million primary school and middle school students, but only 2,700 have the certificate or library card for borrowing books from a municipal library. Shanghai Children's and Youth's Library get annual visits only equivalent to 14 percent of the city's primary school and middle school student population. Another CBBR's survey shows that the quality of collections and services are the main reason for libraries' poor performance. In answering the question, "Do you want to visit a library?" about two-thirds of the students gave a negative answer. "Why?" About 52 percent of students think that the library's collection is limited and old. There is a library building program, identified as the New Century Reading House Program, proposed by the Central Committee of Chinese Communist Youth League that sets a goal of building a collection of 30,000 volumes for the municipal public library for youth and another 10,000 volumes on the regional town level. This may help change the role of the library for the young adult if indeed these plans are realized. A Chinese Young Adult New Century Reading Club was to be launched at the Beijing International Book Fair in May 2002. The Club is sponsored by the Central Committee of Chinese Communist Youth League and offers books at discount prices for young adult readers. Book clubs in China are usually hard to be profitable and there is considerable concern that this new venture will be financially viable.

What Chinese Young Adults Read: Balancing the Recommended and the Chosen

Reducing the burden of required textbook reading has been good news to the promotion of young adult reading. It does not only mean that Chinese young adult readers will have more time for reading, but also means that there will be fewer and fewer supplementary books they have to read after class.[6] Publishers have stated that this change will provide a valuable opportunity in China. Of course there will be an obvious balancing for the young adults between the recommended reading and those chosen at will by themselves.

Surveys show what young adults, who are mainly made up of the middle school students in urban areas, actually do read. In the CBBR Reading Survey of a Middle School in Shijiazhuang, students reported that stories of campus life, literature classics, and psychological self-help titles are their top favorites. There is a category of books that tell stories of students' lives that are written by young authors who sometimes are teenagers themselves. The category of student's life became popular only in recent years, some of the books in the category have become top best selling titles. *Three Fold Gate* published by Writers Press and *Flower Season, Rain Season* published by Haitian Press have climbed on the annual bestseller list of China Book Business Report (numbers 2 and 19, respectively).[7] It seems true that young adults are increasingly developing their self-esteem, which did not exist under educational pressure from family and school. Young adults have found more and more ways to stand up for themselves as Chinese families are getting more and more tolerant in defining a future for their children. They often behave over enthusiastically when attending young adult TV shows and listen to radio programs in which they take every opportunity to express themselves. They are learning to express social concerns and their real needs. It is generally believed that the stories written by young adults about their lives have contributed much to the success of the earlier mentioned two books as well as other titles. It seems that Chinese publishers have learned from the previous successes and are trying to develop more young adult authors and their works. Best selling young adult titles nationwide also include those that describe personal success. A book called *Harvard Girl Liu Yiting: A Success Story* published by Writer's Press climbed to the top best selling list in April 2000.[8] The book is an amazing story of a Chinese student who went to Harvard directly after completing middle school. Of course, the readers of this book are not only young adults who want to learn from the girl's experience, they are also parents eager to read about a model for their children's academic success.

The second best loved and most read category of books is literature classics. Although I found this reported trend to be questionable, the literature classics include novels and poems of old times. These books were written by Chinese authors and foreign authors. Chinese traditional works include such well-known four classic novels as *The Dream of Red Mansion, The Story of Rebels by Water Sides, The Story of Three Kingdoms*, and *The Travel to the West*. There are Chinese families that seriously recommend and help their kids read the four classics because they are considered to be the very first steps in developing a Chinese cultural background for any young person. However, other works that are reported and appeared on the favorite list might not be as appropriate. *The Making of a Hero*, a Soviet semi-autobiography of a soldier, is an example for a translated volume. The book that tells a lot about revolutionary spirits was extremely popular in the 1950s and 1960s when China and the ex-Soviet Union had an alliance and the story is old-fashioned in every way. There is no obvious reason for the energetic and curious Chinese young adults to like reading such a book. One interesting thing I found in the CBBR survey is that students too often mixed up the authors and the books within this category. It is safe to say that students do not really like to read them, and the simple reason for them to list them as their favorites is to gain the approval of teachers. There is actually a strong influence based on the recommendation that teachers and the government make for reading lists in this classical literature category. The CBBR survey had tried its best to avoid this usual bias, but like many surveys, it is impossible to carry out a survey without the help of teachers and schools that try their best to show the "education" approved reading lists and the directly related effects. Both the government and schools have found out that students read less than before and the reform of language teaching, which has a lot to do with reading, has been an important current topic. A 1999 survey done by the Beijing Normal University to examine the reading situation for the reform of Chinese language teaching shows the same trend in which 39.7 percent of students have read *The Story of Rebel by Water Sides*, 33.7 percent have read *The Travel to the West*, 34.3 percent have read *The Story of Three Kingdoms,* and 30.1 percent have read *The Dream of Red Mansion.*[9] The reform of Chinese language teaching started because of the fact that Chinese language teaching provides 300 to 400 articles to read in six years of secondary education. In March 2000, the Ministry of Education approved the new Standard of Chinese Language Teaching for the Nine-Year Compulsory Education, in which reading promotion has a major emphasis. According to the new Standard, Chinese junior middle school students will read no less than 800,000 Chinese characters in a school year.[10] The new Standard issued later a list of recommended after-class reading that contains 100 literary works that represent the traditional emphasis toward reading. A publisher in Liaoning, a province in northeast China, published a complete collection of titles according to the Official List and expects to make a lot of money. The recent strengthened recommendation for middle school students' reading would definitely increase young adults' reading. However, the stronger the reading recommendations in the education system, the more powerful role it gains in education, the less self-determined the young adults' reading gets. It leaves for us an interesting topic to see the balancing of the influential power of teachers who actually read no more young adult titles or books than their students. Some evidence shows that the recommended reading materials focused on education, are different from what the actual young adult readers in the market reported in the survey.[11]

Another new trend we found in the CBBR Reading Survey is that the number three favorite category includes psychological self-help titles. It is reasonable as well as supported by facts that the over-burdened Chinese students have more psychological problems and read more self-help titles. This subject area or topic is getting more and more controversial as parents and teachers have accused some of the titles in the category of teaching kids to be too sophisticated or adult in dealing with human relations. Another notable fact for this category of psychological self-help is that magazines and newspapers play a major role in delivering content. According to the surveys that we collected for this study, magazine and newspaper reading has taken up a considerable portion of reading while book reading remains holding a major position:[12] *Reader's*, the Chinese version of the American *Reader's Digest*; *Youth's Digest, Boys* and *Girls* are all circulated in millions; science fiction, sports, and pop music readings, stories, and articles are all published in magazines and newspapers.

In the 1970s and 1980s, what young people read for pleasure was later a negative category of books—comics. The format of the Chinese comic book is different from the American comic book and the illustrations are mostly black and white with text at the bottom of each page. It is that format of comic books that come to mind when people refer to reading for pleasure. As early as the 1950s, comic books were published in vast number of titles and quantities. It was the reading material most accepted by young readers and what was not mentioned was the fact that for decades these comic books were used for political propaganda purposes. Most of them told a story in one book and some were sold as a series. The content could be anything ranging from stories adapted from the classics to original stories written for this format. The illustration quality varied and some artists were popular and famous amongst the readers. This genre of comic books reached its peak in the early 1980s and in 1983 China published and printed over 0.8 billion copies. A sudden decline took place in this sector of publishing, which then caused a collectors' market to develop with an estimated membership of 20,000 collectors of comic books in China. It is the consensus that the content and format had peaked out and a new trend moved towards the Japanese cartoon books or *manga* became popular. Today, foreign comic books dominate the market although these cause concern due to the sexual, mafia gangs, violence stories as content of some of the Japanese *manga* titles. The age group of readers for the imports is the young adult and no longer the younger age group that read the more traditional Chinese comic books. Therefore, there are serious calls for returning to the old format comic with the more appropriate content for young adults. The China Book Business Report issued a special report on comic books in July 2001 to reflect the call for revival of the traditional comic books based on market research of Chinese and American comic books. However, the Chinese publishers find themselves not able to bring back this sector of comic book publishing after more than ten years of serious review and research. There are still those who consider it to have real potential for bringing back the fun of reading comic books to young adults.

The different research data and summary of several research studies provide several views of the young adult reader and what they are reading in China. It is a very large and complex market segment going through rapid changes. Therefore, this study is a current view to provide an insight and guidelines for a young adult total population of 52,292,606 in China.

Notes

1. A study on reading habits of Chinese young adults is certainly difficult, because there is no authoritative, large-scale survey concentrated on the subject. The China Publishing Research Institute did the only national, comprehensive survey of reading and publication purchasing habits survey in 1999. However, the survey drew samples only from readers ages 18 to 70.
2. China National Bureau of Statistics, Press Release on the Fifth National Census, March 23, 2001.
3. Shijiazhuang is the capital city of Hebei Province; it has a population of 1.3 million.
4. "CBBR Reading Survey of a Middle School in Shijiazhuang," *China Book Business Report*, March 2001.
5. According to the 1999 National Reading Survey by the China Publishing Research Institute, 1998, the average individual's (aged 18 to 70) consumption of newspaper, magazine, and books is 246 RMB yuan (30 U.S. dollars).
6. The new policy from the Ministry of Publication actually ruled that no supplementary books are to be published and distributed through schools. This has caused a certain degree of panic in the Chinese book publishing industry, which soon realized that this may actually open up a huge market as the old market shrinks slowly.
7. "CBBR National Annual Bestselling List/Fiction," *China Book Business Report*, February 27, 2001.
8 "Monthly National Bestseller List/Non-Fiction," *China Book Business Report*, May 29, 2001.
9. Cunrui Middle School Grades Four and Five Reading Survey, by Chinese Language Department, Beijing Normal University, November 1999. The survey listed the top five favorite types of books as follows: detective and mystery stories (46.0 percent), literature classics (45.1 percent), stories on campus life (43.3 percent), knight stories (22.2 percent), and love stories (14.8 percent).

10. A medium-sized novel contains around 200,000 Chinese characters.
11. An interesting survey on Chinese middle-school teacher's reading found that 70 percent of Chinese middle-school teachers read after work less than an hour per day. And in this less than an hour, the teacher's book is their most read type of book, and the better (as regarded by government and school) teachers are, the more they read the teacher's books. *The China Youth Daily*, April 13, 2001.
12. The Cunrui Middle School survey reported that 44.9 percent of students cited novels as their most read books, while prose (mostly in book form) took up 24.8 percent, poetry (mostly in book form) 8.2 percent, newspapers and magazines 21.9 percent, cartoon and comic books 17.4 percent.

8

The Reform of the Book Distribution Industry and the Development of Non-State-Owned Bookstores in China

Yang Deyan

The Book Market of China in 2000

In China, about 7.5 billion copies of domestically published books were sold in 2000, with the total turnover recorded as RMB37.7 billion. These figures indicated respectively an increase by 2.3 percent in units and by 6.2 percent in revenues compared to the figures of prior year. The total number of established bookstores in China reached 76,136 in 2000, of which 14,305 were state-owned and 61,831 were non-state-owned bookstores. Among them 1,066 were specialty bookstores and others are general retail stores. The state-owned Xinhua Bookstore system represented 96 percent of all of the industry's unit sales volume and 85 percent in terms of revenue value. Although the number of non-state-owned bookstores was 4.6 times that of state-owned bookstores, the annual turnover of non-state-owned bookstores was only about 42 percent of that of state-owned bookstores. The year 2000's retail sales of books were mainly achieved in urban areas or 73 percent, while sales in rural areas were only 27 percent of the total. There were 43,011 bookstores in urban areas and 33,125 in rural areas.

The Progress of State-Owned Bookstores' Reform

The Fast Growth of Distributing Groups Co. Ltd.

The reform of the distribution industry in China has experienced several phases since the 1980s. Currently, distribution is primarily represented by the deepened restructuring of state-owned bookstores and the rapid development of a new kind of enterprise, that is, the legal form of the "distributing groups Co. Ltd." Namely, a "distributing group Co. Ltd." is defined as an independent decision-making economic entity, with all members closely integrated in terms of financial control and business operation. Three distributing groups Co. Ltd. with the approval of the central government were first established in 1999. They are designed to take the lead in the reform process for all distributing groups to operate under multiple ownership and gradually across the government's administration boundaries. Such a new orientation of reform serves the overall target of establishing an opened, unified, and competitive national book-market.

Among the three new distributing groups, Guangdong Xinhua Bookstore Group Co. Ltd. is a representative case. It was established on June 29, 1999, a multi-stock owner enterprise with the government agency Xinhua Bookstore in the Guangdong province as the holding company. And other stockholders include ninety-six enterprises, of which are eleven publishing houses and four bookstores from other provinces. Clear objectives are set by this enterprise, that is, to transform the traditional operation mode under a centrally planned system, to establish modern distribution systems and operation modes. The goal was to establish normal chain stores, free-chain stores, and chartered chain stores, as well as to develop gradually coordinated management with the publishing and distributing enterprises in the neighboring provinces. Eventually the new company plans to become the regional book distributing center of southeastern China, including Hong Kong and Macao Special Administration Region. In 1999, this distributing group generated a sales volume of RMB 336 million, 1.3 percent higher than that of 1998. Meanwhile, RMB 25 million of profit was achieved in 1999. When the restructuring was completed by the end of 1999, the gross assets of Guangdong Xinhua Bookstore Group Co. Ltd. registered as high as RMB 307 million, of which RMB178 million was reported as net assets.

Nearly twenty regional distributing groups had been set up by the end of 2000 with the approval of provincial governments, in addition to the aforementioned three national distributing groups. Beifang Book Center of Shenyang City in the Liaoning Province is another good example. The development of various distributing groups Co. Ltd. has contributed greatly to the deepening of the reform of the book distributing system and stimulated the market-oriented development of the publishing industry in China.

Generally speaking, the major concern of reform in the current phase is to transform state-owned bookstores into shareholding enterprises, and the diversification of shareholders is encouraged. Usually, there are three types of stockholders in China that function at this time: (1) employees and staff members of the enterprise; (2) local or non-local, domestic or international enterprises; and (3) government agencies. Varied stock proportion structures by different types of stockholders can be easily observed among all the distributing groups mentioned above, hence the varied modes of management and administration.

Innovative Efforts to Extend Trade Book Distribution

In order to increase book sales and enlarge the trade book market, many state-owned bookstores have established large wholesaling and retailing centers according to the principle of rational macro allocations. These centers often accommodate a variety of books, and are equipped with quick access to information and perfect computer systems. By the end of 2000, nearly all provincial units in China had established such wholesaling and retailing centers in their large cites.

At the same time, many innovative buying and selling approaches had been developed and put into use according to the realities that exist in cities with their own specific market segment conditions. Among all these innovative efforts, three outstanding achievements can be summarized: (1) the popularization and improvement of agency systems, which resulted in the emergence of various types of agencies including the general agent, regional agent, subscription agent as well as storage and transport agent; (2) encouraging the development of chain stores, including the normal chain stores, free-chain stores and chartered chain stores, which have contributed greatly to the accelerated development of bookstore networks; (3) stimulating the development of Reader's Clubs. Presently, many bookstores and publishing houses have Reader's Clubs with their own characteristics, and offer individualized services for their registered readers.

The Development of Non-State-Owned Bookstores

The Current Status of Non-State-Owned Bookstores

After more than twenty years' development under reform policy, a new pattern of book distribution system has taken shape in China.

The state-owned bookstores remain the main part of the system, while they are complemented by numerous vibrant bookstores with different types of ownership.

With the development of market economy in the last twenty years, the non-state-owned bookstores have already become an important component of China's book market. The historical growth of the non-state-owned bookstores is reflected in various aspects, such as the enlargement of scale, the changing status from supplementary to independent development, and the management improvement from being informal to being formal or more professional. The number of non-state-owned bookstores rose to some 60,000 in 2000, of which a large number were private book stands. And among 250,000 people engaged in book distributing, more than 100,000 people are employed in non-state-owned bookstores. Peking University Press presented awards to thirty top bookstores in terms of sales volume in 2000, and twenty-eight winners were non-state-owned bookstores.

The Books and Periodicals Distribution Association of Beijing and Beijing Press Service Center jointly held a book fair especially for the non-state-owned bookstores during the 2001 Beijing Spring Book Fair. It was a milestone in the development of non-state-owned bookstores in China. More than 1,500 non-state-owned bookstores and over 200 publishing houses of books, periodicals, and video products participated in the fair. The number of titles exhibited surpassed 7,000.

The Development Phases of Non-State-Owned Bookstores

1. The Start-up Phase (1982–1987). Non-state-owned or independent bookstores emerged in the late 1970s and early 1980s only as supplementary outlets to the Xinhua Bookstores, the national system of state-owned bookstores. Since 1986 when the existence of non-state-owned bookstores was officially approved, the bookstores of collective ownership and private ownership boomed all over the country. In 1987, the number of collective-owned and private-owned bookstores was recorded as 10,814, which was 1.18 times that of the state-owned bookstores, while the total number of all kinds of non-state-owned bookstores was 9,194 including bookstands and bookstalls, which was 10.75 times that of the state-owned bookstores.

2. The Development Phase (1988–1991). Although still in the primitive development period, non-state-owned bookstores became strong during this phase. A number of non-state-owned bookstores with economic strength and business success had grown up in the competitive

market, and became well recognized throughout the whole country. Good examples of these successful bookstores are Feng-ru-song (Forestsong) Bookstore and Guo-lin-feng Bookstore in Beijing, Ji-feng Bookstore in Shanghai, Xue-er-you Bookstore in Guangzhou, Zhi-yuan Bookstore in Jinan, and Xi-xi-fu Bookstore in Guizhou.

3. The Improvement Phase (since 1992). In this phase, the non-state-owned bookstores had already taken the form of large-scale operations; they began to develop cooperative management, and played an important role in the country's book market. In 1999, there were 35,282 collective-owned and private-owned bookstores. The total would be even larger if we included all other kinds of non-state-owned bookstores. In Beijing, the number of non-state-owned bookstores was 13 times that of all state-owned bookstores. And in 2000, the number of non-state-owned bookstores in overall China was 4 times that of the state-owned.

The Difference of Management Compared to State-Owned Bookstores

1. Non-state-owned bookstores have no general wholesaling right like state-owned bookstores. However, non-state-owned bookstores have attained similar authority like state-owned bookstores on the wholesale markets at the provincial level and below.

2. State-owned bookstores have the monopoly and authority to engage in school textbooks sales, especially the textbooks for primary schools as well as junior and senior secondary schools.

The Types of Non-State-Owned Bookstores

1. All non-state-owned bookstores can be divided into four types according to their function. (A) Retailing enterprises including bookstores, bookstands and bookstalls. The number of this type of accounts represents about 80 percent of the total non-state-owned bookstores in the country. Most of them are in the state of being self-sufficient. (B) Wholesaling enterprises, usually with rather limited business scale and size. The annual turnover of each enterprise of this type is between RMB 2 to 5 million. (C) Chain stores, which are the duplicated extension of some bookstores with high economic strength. (D) Internet bookstores, which are still in the experimental development stage.

2. All non-state-owned bookstores can be divided into three types according to the source of capital and the initial ownership structure. (A) Private enterprises that are started on the basis of self-generated in-

vestment funds. This type of bookstore accounts for over 90 percent of the total non-state-owned bookstores. (B) Enterprises with mixed ownership. The ownership of bookstores belonging to this type is complicated because they tend to be a combination of state, collective, and private property investments. Former state-owned bookstores, which were contracted out or transformed to private forms, fall into the category. (C) Innovative enterprises, which are mainly based on capital input from other industries. The dynamic vitality of this type of enterprise is well exemplified by the recent development of Internet bookstores.

The Problems Facing Non-State-Owned Bookstores

The major problems that face non-state-owned bookstores are due primarily to their unfair market competition with state-owned bookstores. On the one hand, non-state-owned bookstores are not endowed with general wholesaling rights like their state-owned counterparts. On the other hand, some non-state-owned bookstores still have to cope with informal and problematic market conditions, for example, allowing very high discount rates, which affect the normal operations of the book market.

The Books and Periodicals Distribution Association of China was established in March of 1991. Currently, it has 3,650 members from state-owned and non-state-owned enterprises, and consists of fourteen working committees, including the working committee of non-state-owned bookstores. It can be expected that with the deepening of the reform in China, this organization will take over additional functions that have been traditionally handled by the government agencies. The Association will also play increasingly important roles in the development of an open, professional and healthy book market in the country.

In December of 2001, China joined the WTO. In the following year, China allowed foreign-service providers to run book, periodical, and newspaper retailing businesses in five economic special zones of Shenzhen, Zhuhai, Shantou, Xiamen, and Hainan, and in six cities of Beijing, Shanghai, Tianjin, Guangzhou, Dalian, and Qingdao. Four joint venture retailing businesses will be allowed to open in Beijing and Shanghai, respectively, and two in every other city. After joining the WTO, Zhengzhou and Wuhan retailing markets will be opened to joint venture retailers, and in two years, all of the provincial capital cities and the cities of Chongqing and Ningbo will also be opened. Three years after China's joining the WTO, foreign-service providers will be allowed to distribute books, periodicals, and newspapers in China. It is the general belief that as a result of China joining the WTO, the country's book distribution industry has been enhanced by the communication, cooperation, and competition with foreign counterparts, and it has provided China with a great opportunity in its continuing development of this industry.

9

Economics of the Chinese Book Market

Sun Qingguo

The bestseller, a ghost loved and hated by the book business, is about to prevail in China. For decades, Chinese publishers tried their best to create long-standing titles, or "longsellers," rather than bestsellers. Bestsellers were considered short-lived, low-class, and lacking the higher image that "the book" has always represented in China. But now the practice of best-selling titles is here in China, and publishers who struggled for years with their financial well-being are actually beginning to benefit from bestsellers.

Major Changes in the Chinese Book Market: Late 1970s to the Present

Since the late 1970s when China started to open and reform its economic systems, the Chinese book industry went through four periods: (1) low title output/underserved market, (2) buyer's market/ increased title output, (3) segmented market, and (4) bestsellers.

The Underserved Market Period When Any Book Would Succeed

After ten years of social turmoil (The Cultural Revolution, 1966–1976), the Chinese book publishing market shrank to an extreme degree. In the late 1970s, Chinese readers lined up in front of bookstores waiting for books. Practically any book available was welcomed and would succeed. Books of all categories were sold, and most titles sold vast numbers of copies (millions). Nevertheless, it is not right to say that these books were bestsellers. As well, the number of titles in print was small despite a popular slogan in the publishing industry at the time: "Publishing More Books and Better Books." After ten years of effort (1970 to the mid 1980s), however, the shortage of new titles was overcome and a considerable amount of new books satisfied the basic reading needs of the Chinese people.

The Buyer's Market Period When Title Numbers Exploded

As market needs were more and more satisfied, the number of copies sold of a single title decreased dramatically. Publishers, who needed to achieve their financial goals, began to publish more and more new books in terms of title numbers. During the transition from the 1980s to the 1990s, a national title output of over 100,000 was achieved and the increasing trend continued. Then the government changed its policy: "Control Title Output, Care More about Quality." However, even with the strictest government control to limit licensing ISBN book numbers, the number of new books grew, even geometrically.

In early 1997, I witnessed the opening day of a major bookstore in a capital city of a province. The day's total sale was over 1,000 books, with a total sales volume of 20,000 RMB Yuan. However, the unit sales of 1,000 involved roughly the same number of titles. The title with sales of more than two copies was only a dozen in number. Those titles with more sales were textbook supplementary readings for students. This phenomenon told the booksellers that bookselling is "title selling," the more different titles on the shelf, the better the sales. This strategy was common sense and the prevailing business practice among book retailers. The booksellers' demand for more new frontlist titles was then a clear signal to publishers. At all ordering conferences, publishers tried their best to promote frontlist titles, and some of them released hundreds of new frontlist books. More new bookstores with much larger floor space again resulted in an increased need for adding new titles to the shelves. During that time, bookstores were competing for size (both the square meters of floor space and the number of new titles on the shelves). Publishers enjoyed this very much since all of their new titles could be displayed, and more books could therefore be published.

In this explosive title growing period, however, statistics from the government show that the market (from the use of paper) did not grow as much as compared to 1987. This means the average copy sales decreased greatly. Publishers began to hope that a certain number of their titles would gain more print runs, though they did not have any idea or understanding of bestselling titles at that time.

Year	Number of Titles
1995	101,381
1996	112,813
1997	120,813
1998	130,613
1999	141,831
Compound growth 1995–1999	40,018
	39.5 %

Source: China Book Business Report.

Market Segmentation and Adjusting

When the average copy sale dropped greatly, the industry believed that it was just temporary and expected that the single title ordering would again reach previous levels. However, the market did not act as the publishing industry expected, but instead headed in another direction. The reader's tendency in selecting books became more obvious, and many traditional formats of books were going through a social change. Some categories of books became less popular, some even disappeared. Literature works, which used to be the most popular category, were read less. Some noted that, ironically, the people who write poetry outnumber those who actually read it. The Chinese comic book (in its traditional format) and picture calendars disappeared. Suddenly there was a great contrast to the previous peak years of success. In their peak years, comic books sold 0.8 billion copies and picture calendars sold 0.7 billion copies. A group of specialized publishers started and became very popular. But these publishers found themselves in a situation where people stopped reading what they published; and they are still wondering how to find new markets. At the same time, English-language learning, business and computer books became new subject areas in the market. Publishers in these fields helped the segmentation of the Chinese book market. They quickly fostered market growth, publishing new types of books that readers needed. Since 1999, the Chinese book industry has grown significantly each year. However, not all categories have grown; some have actually shrunk. Interestingly, only a select number of Chinese publishers contributed to this growth.

The Book as a Form of Popular Entertainment and the Start of Bestsellers

The book, as traditionally understood by the Chinese people, is not a product that disappears. Today, however, the role of the book in the Chinese market is different. Numerous titles and authors have made "the book" less respected, and it has lost its intellectual image. A book's lifespan has suddenly become short. What sold yesterday does not necessarily sell today. Readers' needs and interests have become difficult to identify and measure. Identifying a solid, respected list of good books has become an increasingly difficult challenge. The book in China, in a sense, has become an ordinary kind of consumer product.

The books that readers consume, usually for one-time use only, satisfy entertainment and relaxation needs, especially in so-called commercial society. Most books entertain and have a short life cycle. The revision and updating of information is much faster today, which also contributes to this shorter life cycle. As well, readers' consumer behavior and values change. One book becomes everybody's book, and once fashion changes, it becomes nobody's book. Hence, the arrival of the concept and practice of the bestseller in China.

Shares: How Bestsellers Contribute to the Market

In the beginning people would only see the coming and going of bestselling books, without statistics and judgment of their market values. This practice changed, and Chinese publishers began to pursue market values in relation to what they published. Publishers found themselves asking the questions: What is the relationship between the number of titles published and revenue? Does more titles mean more revenue? The better Chinese publishers then sought to publish books that could accomplish two goals: (1) attain bestselling status, and (2) possess lasting quality. As a whole, they now care more about the release of bestsellers in the market.

According to the Beijing Open Book Market Research Institute's statistics from the National Book Retail Tracking System, Chinese book publishing has entered into an era of bestsellers. What are bestsellers in China? Of course fiction and non-fiction, but also books that fall into the specialized sector, where a small number of titles enjoy a majority of the market share.

It should be noted that in 2001 the National Book Retail Tracking System monitored over 480,000 titles that sold at retail channels.

What was cited as the rule of 2 to 8 is clearly supported by the tables. This rule indicates that publishers concentrated on trying to publish the 20 percent titles with 80 percent market share. To publish more titles as a strategy for growth no longer worked for the publishers, and booksellers no longer believed that more titles meant more sales. All these elements influence the mechanism of bestsellers in the market.

The Top Bestseller Myths in Recent Years

Chinese publishers used to publish books that would create positive social response, not to mention their ideals of publishing classics that would last centuries or forever. Many cultural critics have pointed out that it is not always good literature that sells best. However, as we go through the bestsellers in recent years, it is noted that most of these titles reflect current times, and are a result of both publishers' and readers' interactivity in the changing market.

Learning, Education, and Personal Success

Since 1999, most of China's big bestsellers have been books with an education theme. *The Learning Revolution*, *Quality Education in the U.S.*, and *Harvard Girl: Liu Yiting and Her Quality Education Story* have all topped the national bestsellers' list. When the Clever Company heavily promoted *The Learning Revolution*, the book industry was shocked that this type of book could be marketed so successfully. The Clever Company, which is an educational software company, is still proud of this achievement. Although the so-called 10 million print run was an exaggerated number, it achieved record sales according to the Beijing Open Book Market Research Institute. It is ridiculous for many publishers trying to find out whether the Clever Company has benefited from "The Learning Revolution," especially in today's economy. It is pioneering for the industry that this publisher used modern marketing campaigns in the book market and it aroused the imagination of many to understand the making of a bestseller.

Too many people were misguided by the media in promoting *The Learning Revolution.* It should be noted that learning is a subject matter that is very important for many Chinese people. And as to the *Quality Education in the U.S.* and *Harvard Girl*, we not only see the "myths" of the talented students but also see the heart of the overall education process and system. People expect to change their social positions through learning; parents invest in education for a better future of their child in our present society.

Fortune, Finance, and Financial Quotient (FQ)

There are many hot social topics that provide the opportunity for a bestseller. Economics, money, and investing have captured the attention of many readers. From Samuelson's textbook *Economics* to the *Harvard Business School MBA Textbook Series* and *Stock Investing: Just a Few Skills*, Chinese stock dreams are nurtured by such bestselling titles. Though a large number of Chinese have recently acquired wealth, there is an even greater number of Chinese who want to be rich. *Rich Dad and Poor Dad* is another series that became popular and pushed the book market to grow in this subject category.

Money and fortune have never been on par with the traditional Chinese values of high ideals, goals and happiness [though the government does call for its people to become successful and make their country stronger]. However, people *are* reading these bestsellers, which has never happened before.

Celebrity Publishing

In our commercial world, "Celebrity" publishing is a phenomenon. In recent years, from Zhao Zhongxiang (CCTV host) to Cui Yongyuan (CCTV host), famous TV celebrities have published books. When Zhao Zhongxiang published his second book and received a weak market response, critics suggested the celebrity publishing game was over. However, with other celebrities joining in, the sector is as hot as ever.

Their appearance on TV is perfect promotion for their books. *Just so so* by Cui Yongyuan started to sell in July 2001 and by the end of the year, 1.3 million copies were sold. Generally, celebrity books are easy to read. They're like a kind of "cultural fast food" easily accessible to the mass consumer market. Books by Yu Qiuyu (a Shanghai scholar) and Liu Yong (Taiwan writer) functioned in much the same way. Their personal "legends" are quite ordinary, but in a culture and society where everyone is ordinary, myths and legends have to come from the "ordinary" celebrities. Foreign celebrities such as Jack Welch also are a hit in the Chinese book market.

Movie and TV Drama Spin-offs

Novels have been falling in popularity and sales for years. Biographies and prose have fared better. Author Yu Qiuyu writes prose and many of his books make the bestseller list. Writer Han Han, however, was able to achieve bestseller status with his novel *Three-fold Gate*, which makes it difficult for publishers to understand their market. However, since the start of 2000, many novels have begun to rebound and take the market away from prose. Most of the novels are related to movie and TV dramas. When a movie or TV drama gains popularity, the book version also achieves popular status. *Yongzheng Emperor*, *Come and Go*, *Good Morning, Miss*, and *Hand in Hand* are all bestsellers based on TV dramas or movies that received solid market response.

Pure literature and serious writers are in a sad state. This category of publishing is not very close to the mass market and, therefore, will reach only a small segment of readers in China. The fiction market in 2001, however, did improve thanks to a few bestsellers, although big fiction titles with sales of over 1 million copies have yet to be seen.

The Powerful Translation Titles

Several years ago, people said foreign novels would never succeed in China. Comprising the fiction list, except for Hemingway's *Old Man and the Sea* and another Japanese writer's work, are few foreign titles. Interestingly, the non-fiction list bears a sharp contrast, with translated works heavily represented. As noted earlier, books on finance, computers, and English-language learning make up this list. Translated works are also very popular in the children's category. The *Harry Potter* books are one example, but the overall children's book category includes many translations.

The appearance of bestsellers exemplifies the acceleration of the Chinese book industry's step toward the consumer market and the concept of marketing. Bestsellers are the product of the steps taken towards advertising, promotion, and publicity for this market segment. Bestsellers have also helped push the market to grow. A publishing house's management for a bestseller is an important indicator of its ability for building brands and marketing.

How Bestsellers Change the Future Marketplace

Today, more and more people are looking at bestseller titles. Publishers, authors, booksellers, and readers are paying attention to current bestseller lists, which is a new development in China. Authoritative bestseller lists have been and will continue to be the indicator of the industry. It guides the industry to adjust the structure of printing quantities, ordering, sales, and inventory management. Publishers are examining their responsibilities and related activities in influencing the readers' and booksellers' choice alike. Indeed, the bestseller will influence and change the structure of the Chinese book market.

Bookstores Will Concentrate on Trade Books, Especially Bestsellers

With the rapid and wide implementation of computer technologies, readers will look for and buy specialized books through online bookstores, mail orders, and research engines. Bookstores in China will be concentrating on selling trade books and providing services to a large numbers of readers. Bestsellers will take up a large proportion of their sales. Large bookstores will not continue to develop because more and more new bookstores will be specialized.

Brand Strategy Will be an Emphasis for Competition

Although China's publishing houses are all specialized in a field or subject area, most of them have developed their own trade book programs. There are academic publishers in the real sense in China, but they have added a trade publishing program or list to their core business. Therefore, the market share is important in deciding a publisher's fame and position in the industry. And bestsellers effectively decide the market share. If a publishing house does not have a bestseller, it has no brand name or identity in the industry. Over the past few years, Chinese publishing houses have been pursuing a minimum goal of having their books displayed in a few government-managed bookstores. But in the future, publishing houses with no brand recognition or real marketing capability will have difficulty getting their books displayed in the growing number of private enterprise bookstores. Booksellers will not put too many slow selling titles on their shelves (although the best selling titles are still limited in numbers).

Title Numbers Will Relatively Decrease

Publishing houses are used to trying to enlarge their lists of new books, but bestsellers will change this situation. The market and related distribution system in this huge country will counteract this trend. Long lists with no bestsellers will make a publishing house less competitive. To enlarge lists by producing more and more new titles will cause waste and loss of investments. Publishers will have to publish fewer titles and maintain a more focused effort than they have in the past. It is of course true that not all books have the chance of becoming a bestseller, and of course bookstores will not sell only bestsellers.

On the up-side, the Chinese publishing industry will also curb the publishing of low-quality books, and Chinese readers will applaud and welcome more quality books in the market.

About the Open Book Market Consulting Center

The Beijing Open Book Market Consulting Center is the first and only book market research center in China, and it is similar to the British BookTrack under Whitaker and BookScan in the United States. The Beijing Open Book Market Consulting Center, along with the China Book Business Report, established a National Book Retail Market Tracking System to collect monthly entire sales records from POS in nearly 100 major bookstores in fifty major Chinese cities. The System covers 12 percent of the whole Chinese book retail market, and monitors all titles published in the Chinese language. In 2001, the Institute's Books in Print list contained over 480,000 titles and monitored the market performance of all book publishers in China.

Beijing Open Book Market Consulting Center scientifically analyzes book sales records and reports promptly and accurately to the industry to help publishing professionals understand the market better, its development, trends, and the competition, and forecasts.

Beijing Open Book Market Consulting Center's pioneering work fosters Chinese book publishing to a market-oriented model, and helps some of the most competitive publishers develop their strategies with its professional information services.

Bestsellers' Market Share in Eight Categories (January-December 2001)

Table 1
Fiction

Title	Share of Total Fiction Book Retailing Sales	Share of Total Retailing Fiction Titles
Top 30	10.03%	0.05%
Top 100	16.09%	0.18%
Top 200	21.61%	0.36%
Top 1000	41.88%	1.80%
Top 5000	70.36%	8.99%

Table 2
Children's Books

Title	Share of Total Children's Book Retailing Sales	Share of Total Retailing Children's Titles
Top 100	18.38%	0.29%
Top 500	33.78%	1.44%
Top 1000	43.61%	2.88%
Top 2000	58.72%	5.77%
Top 3000	67.50%	8.65%

Table 3
Textbook Supplementary Books

Title	Share of Total Textbook Supplementary Book Retailing Sales	Share of Total Retailing Textbook Supplementary Books Titles
Top 100	7.10%	0.15%
Top 500	18.89%	0.76%
Top 1000	27.60%	1.53%
Top 2000	39.30%	3.05%
Top 3000	47.86%	4.58%

Table 4
Business and Management

Title	Share of Total Business and Management Book Retailing Sales	Share of Total Retailing Business and Management
Top 100	17.85%	0.29%
Top 500	33.45%	1.46%
Top 1000	43.17%	2.93%
Top 2000	56.05%	5.85%
Top 3000	64.59%	8.78%

Table 5
Law Books

Title	Share of Total Law Book Retailing Sales	Share of Total Retailing Law Titles
Top 100	21.40%	0.90%
Top 500	41.32%	4.51%
Top 1000	57.06%	9.02%
Top 2000	73.94%	18.04%
Top 3000	83.65%	27.07%

Table 6
Computer

Title	Share of Total Computer Book Retailing	Share of Total Retailing SalesComputer Titles
Top 100	9.34%	0.39%
Top 500	22.43%	1.94%
Top 2000	48.78%	7.76%
Top 3000	59.74%	11.64%
Top 5000	74.08%	19.39%

Table 7
Lifestyle

Title	Share of Total Lifestyle Book Retailing Sales	Share of Total Retailing Lifestyle Titles
Top 100	14.63%	0.36%
Top 500	29.38%	1.78%
Top 1000	39.22%	3.56%
Top 2000	51.94%	7.13%
Top 3000	60.38%	10.69%

Table 8
English Language Learning

Title	Share of Total English Language Learning Book Retailing Sales	Share of Total Retailing English Language Learning Titles
Top 100	30.04%	0.54%
Top 500	49.72%	2.68%
Top 1000	61.70%	5.36%
Top 2000	74.74%	10.72%
Top 3000	82.13%	16.08%

10

China: An Open Land for the Rights Business

Hu Shouwen

First of all, allow me to introduce China Youth Press, which is one of the largest and most influential presses in China and has been in business for fifty years. China Youth Press publishes 700 titles annually under five different imprints, and of this total about 60 percent are new titles. Starting in 1996, the publisher started buying a large number of translation rights from U.S. publishers. Early in 1998, *Publishers Weekly* sent two correspondents to Beijing to interview several publishing professionals, including myself as the president and publisher of China Youth Press. This was the first time *Publishers Weekly* released a full report in a special edition on the new generation of Chinese publishers. Another key point that needs to be made in this brief introduction is that three chief rules govern the publishing industry. The 1997 Regulations on Publication management, the 1997 Regulations on Management of the Printing Industry, and the 1998 Regulations on Electronic Publishing stipulate that only government designated publishing organizations may engage in publishing. Therefore, it is not possible for private individuals or foreign entities to engage directly in publishing in China. Foreign companies or organizations are also prohibited from owning distribution companies or taking the majority share in joint ventures. Another key factor is that English-language publications have an extremely limited appeal. The foreign language abilities within the broad consumer market and the import controls on foreign language publications limit the size of the potential market for English-language books and magazines. As a result, reprint, translation, and co-publishing rights take an especially important role in the publishing industry of China.

Now, I would like to help colleagues in the Western world understand the status and increasing business opportunities of the Chinese publishing industry within a huge and growing market. Time Inc. published a special report on "Young China" in the October 23, 2000 issue of *Time* magazine. The population of the country was documented in the following terms: the total population is 1.2 billion, of which 630.5 million are under the age of 24. In contrast, the combined population of North America, Australia and Russia is 622 million. The number of university students in China was 6.1 million while in the United States there were 12.6 million in 1997.

Student Enrollments in All Levels of Education (million)

	Elementary	Middle School	College
1990	122.4	51.0	2.06
1999	135.4	80.0	7.13

Source: *China Statistical Yearbook 2000*

According to the Press and Publications Administration, there are about 560 publishing houses in Mainland China, which is the fifth largest book market in the world. Beijing is the center of publishing and media in China; there are over 200 publishers in Beijing. Some of these are China's largest publishing houses with long histories. China publishes 140,000 new titles each year, of which 6 percent are translated works. Some people claim that China is number one in the world in terms of titles published annually, but the revenue sales figures and profits are far lower in comparison to other countries. Many literary agents and publishers in Western countries do not fully understand the economics, average personal income, and list prices of books in China when they request advance payments for translation rights.

Before the 1980s, Xinhua Bookstore, a state-owned national bookseller chain, monopolized book sales. With the economic reforms in the mid-1980s, changes began to take place in the country's book distribution system in that the first non-state-owned book distributors were established. By now, they have grown strong enough to be able to share the total annual sales volume with the former national distribution giant Xinhua bookstores. With its steady growth and development, non-state-owned distributors have experienced continued and steady growth. These positive developments have contributed to propelling the overall reform of the book distribution system in China. The wholesale and retail book distribution channels

Year	Titles	Sales US$ dollars (million)
1995	101,381	$ 1,838
1996	112,813	$ 2,726
1997	120,106	$ 3,562
1998	130,613	$ 4,193
1999	141,831	$ 4,282
Compound	40,018	$ 2,444
growth 1995–99	39.5 %	132.9 %

Source: *China Book Business Report*

will provide foundations for future development of the Chinese publishing industry. Non-state-owned distributors have become an important factor in promoting the translation and reprint rights business.

In contrast to the book publishing industry, the number of domestic Chinese periodicals publishing with a national distribution has remained in the range of about 7,000 titles in recent years. There are a further estimated 5,000 magazine titles with limited local or regional circulation. Therefore, the total number of periodical titles is about 7,000 for China. The top three magazines in relation to circulation are *Fortnightly Chat* with a 6.4 million circulation, *Peasant Digest* with 4.4 million, and *Guangdong Second Classroom* with 3.2 million circulation.

Very few rights transactions were seen in Mainland China before 1986. The first Beijing Book Fair, held in 1986, marked the official beginning of the rights business between Chinese publishers and their foreign counterparts. The last fifteen years can be divided into three periods to better understand the rapid development of the bookselling and publishing industries. During the first period (1986-1990), Chinese publishers focused mainly on selling licenses to Japan, Hong Kong, and Taiwan. The purpose of doing so was obvious—to earn hard currency. During the second period (1990-1997), Chinese publishers were not only interested in selling licenses to various countries and regions but also in buying rights. The number of titles bought and sold in this period was almost the same. During the third period (after 1997), the number of foreign rights bought increased tremendously and for the first time far outnumbered the titles sold. In 1999, Chinese publishers bought 6,461 titles worldwide; this total included 1,080 titles from Taiwan, Hong Kong, and Macao. The majority of titles, 5,370, were from Western countries. The United States ranks first in selling rights to China with a total of 1,920 titles sold, followed by Britain (860 titles), Germany (389 titles), and France (219 titles). While in the same year China only sold 418 titles. Most of foreign rights bought are in the following categories: computer science, economic management, foreign language teaching and learning materials and dictionaries. Titles bought in the form of book series and library editions accounted for 80 percent of the total.

The publishers who represented the strongest buying and selling of rights are based in Beijing, Shanghai, Liaoning, Jiangsu, and Guangxi. When it comes to buying foreign rights, countries like the United States, Britain, Germany, France, and Japan are the prime sources of new publications. As the publishing industry develops a better international understanding, more book rights will be purchased from other parts of the world in order to provide our readers with more meaningful and diversified books from different cultures and regions. The 1999 figure shows that Chinese publishers bought foreign rights from over forty countries, regions, and organizations including India, Malaysia, Mexico, and Peru. The rights business had been very brisk between the mainland and Taiwan and Hong Kong in the 1980s and early 1990s, but then it began to drop. The rights business picked up momentum across the straits in 1998; that year Mainland China bought 870 titles from Taiwan and 120 from Hong Kong. Last year the figure was 700 titles from Taiwan and 310 titles from Hong Kong. In the meantime, the role of Taiwan and Hong Kong as intermediaries in the rights business has been strengthened. The bestseller *Learning Revolution*, which sold approximately 8.5 million copies, was introduced with the help of the Hong Kong-based publisher.

Mainland China is a developing country in terms of economic development. The list price for a single title has increased 17 percent compared with fifteen years ago. A paperback book with 140 pages is usually priced at 8 Yuan, about US$1. Most books are bound paperbacks and range in price from one to no more than four U.S. dollars to meet the limited purchasing power of the public. The average print run of a single title has decreased sharply compared with that of fifteen years ago. For example, the China Youth Press' experiences report that a print run of a single title was 320,000 copies seventeen years ago, while today it has dropped to an average of 38,000 copies per title. Given the fact that China produces 140,000 titles per year, the counterpart is that the average press run per title has been reduced considerably. For most books the initial print run will not exceed 10,000 copies. That is why publishers in China are willing to pay high prices only for bestsellers while the regular trade or other book rights are based on the low list price and related dramatically reduced initial printing. The focus is on the translation rights of bestseller titles listed in American media like the *New York Times*, Amazon.com, and other sources.

In the foreseeable future two kinds of books will become popular in China as its economy continues to grow, per capita income increases steadily, and its long traditions focus on education. One category is the illustrated book, including children's books and adult illustrated books such as art publications, art history, techniques and designing. Chinese publishers are paying special attention to art titles by big name publishers in the West. E. H. Gombrich's *The Story of Art,* published by Phaidon, was published by two Chinese publishers during the past twenty years and sold very well, and has become the must-buy textbook for Chinese art students. The second category includes reference and non-fiction, especially computer, books. For example, *The Road Ahead* by Microsoft chairman and CEO Bill Gates was a success story. In cooperation with Beijing University Press, the American publisher and author have received large royalty payments. China Youth Press just closed an agreement in April 2000, in which the translation rights were acquired for over seventy computer titles at one time. This showed the Press' financial ability in relation to a potential market. It is possible for Chinese and Western publishers to enter into win-win agreements. Although most books do not generate huge incomes for Western publishers, there have been after all a number of successful titles translated and published in the past few years. This is where we want to use our experiences to reach new goals.

China entered the "Berne Convention for the Protection of Literary and Artistic Works" and "Universal Copyright Convention" in 1992. Copyright awareness and adherence have been greatly increased among Chinese publishers. It was especially so when China enacted and implemented a new national copyright law to protect intellectual property rights in China. However, as in any other country in the world, there are always some people who disobey the laws;

China's publishing industry is constantly frustrated by the infringements of copyright, pirated materials, and other illegal activities. One of the measures taken was to establish a national anti-piracy coalition; hopefully it will put a curb on piracy. Chinese publishing houses, though owned by the state, have to face the realities of the market and be responsible for their own profit and loss. Piracy not only harms the interests of our foreign partners but also poses a threat to our own survival. It is an established reality that no Chinese publishers can ignore piracy activities. A Chinese proverb best describes the situation: "Chinese publishers and their foreign partners are two grasshoppers tied together with one cord." We are in the same boat and have the same interests to protect.

American publishers and their agents usually expect quite high advances. To best protect their own rights and interests, most of them, if not all, request a one-time royalty payment for the first printing as an advance. Publishers from Europe and other regions are more flexible and sell translation rights according to the norms of an advance against standard royalty terms for first, second, and subsequent printings. Now that the publishing industry has established a five-year track record in China, it would seem appropriate that translation rights should be negotiated and sold in accordance with international norms. There is fierce competition among Chinese publishers regarding buying foreign rights. High advance payments and royalty rates were and still are a big issue. The royalty rates with related advances asked by Western publishers range from 5 to 10 percent, some as high as 12 percent. In China, authors are paid royalties according to print run, rather than actual copies sold. Therefore, the 10 percent or higher royalty rate for the first printing is too much for the average publisher. The Chinese publishers reached a consensus last year, and, to be more specific, the royalty rate for translation rights should be within 8 percent. In our experience we found that German publishers take a more flexible and more reasonable approach when selling Chinese translation rights. British publishers have presented some obstacles when selling rights for illustrated books. It is often the case that when a Chinese publisher shows interest in a title, the British publisher insists on maintaining control of the printing and binding of the book, and then sells the finished book to the Chinese publisher. This practice has transformed Chinese publishers into their sales agents and "distributors" in China. We find it very difficult to accept this method of cooperation. We understand that the purchase of the translation rights also requires the purchase of printed and bound books to provide the original publisher with the added income of selling the books. The cost of manufacturing books outside of China and importing finished books only increase the average list price and decrease the sales potential within China.

Chinese publishers have their own market requirements in contrast to foreign publishers in comparing the prices and supervising the production quality of printers. We may find other approaches to address the issues of purchasing translation rights that require printing and binding of art or illustrated books. What we really care about is how to reduce printing and shipping costs, which are issues foreign publishers do not always understand in the hectic climate of the Bologna Book Fair or the Frankfurt Book Fair.

The increase in participation of international publishers in the Beijing Book Fair and in exchange the ever-increasing participation of China's publishers in the many international book fairs provides many opportunities for a better understanding of the different markets with related economics. As publishers we are in the communications and information industry. Therefore, we will be able to enlighten each other as we will identify as well as respect the differences to increase the purchase and sale of translation rights.

In recent years, the Western and Hong Kong publishing companies began to focus on the magazine publishing industry in China and started to look for partners in view of the limited number of ISSN with related government approval for publication. At the beginning of the 1990s, the magazine licensing of *Shanghai* and *Elle*, *Disney* and Renmin Post and Telecommunications publications were permitted. Then from the mid-1990s the licensing and cooperation of Shanghang and *Cosmopolitan*, the Chinese Youth Publisher Zhuiqiu and *Figaro* of France were permitted. The three fashion magazine publishers in China took the initial leadership in obtaining permission to cooperate with foreign magazine publishers to develop local editions. At that time it was possible to publish the foreign partner's editorial content with 30 percent of local Chinese content for the Chinese edition of the respective magazines. These three magazines started a period of change in the magazine publishing industry. In 2000, many high level and sophisticated magazines were translated, adapted and published for the consumer in China. Within a few years the Chinese consumer and business magazines joined the world at the same level and standard in content, production and advertising.

References

Buckwalter, Charles. *China's Publishing Industry*, Washington, D.C.: U.S.-China Business Council, June 2000.
China Books and Periodicals: www.chinabooks.com
China Business World: www.chinadaily.net
Haynes, Michael. "When China Wakes." *Folio* (June 1998).
McCarthy, Terry. "Lost Generation." *Time* (October 23, 2000).
Sanguo, Cheng. "Chinese Market for Western Publishers," China Book Business Report. Beijing, China, October 2000.
U.S.-China Business Council: www.uschina.org

11

Publishing Education of China Faces the Challenge of Development

Yu Huiming

Introduction

China's publishing industry began to regain its strength in the early 1980s and tremendous advances have been achieved in this industry in the last twenty years. The number of publishing houses had increased to 566 in the year 2000, while there were only 214 in 1981. The industry has a total of 46,390 employees at present. There were 73,898 bookstore retailers at the end of 1999, among which 10,376 belonged to Xinhua Book Store, and over 240,000 employees were engaged in book distribution.[1] Since the government takes the publishing industry under strict control, the number of publishing houses has not increased much. However, the scale or volume of the business has been growing fast throughout recent years. In 1999, more than 140,000 titles were published in China, and 7,329 million volumes were sold, with total sales estimated to reach a total of $4.3 billion. This means that an average of 5.82 books were sold per person in China 1999.[2]

The most distinctive point in China's book market is that students are the major consumers within the market. According to statistics, textbooks and supplement books account for only 20 percent of the titles published, while their sales volume occupies 70 percent of the whole book market. Textbooks represent the largest amount of the profit of the publishing business.[3] The profit rate of the publishing industry kept increasing by 20 percent from 1996 to 1998 and the profit was $382 million in 1998. It ranked twelfth among the thirty-seven industries throughout the country. The rapid development and growth of the industry included a tremendous recruitment of new employees. The industry has a great demand for employees with related professional backgrounds and training. In addition, with the employment of computers, the Internet, and the introduction of electronic publishing, publishing employees have had to update their knowledge frequently to meet more and more challenges within their own business. Therefore, various educational institutions have had to assume the responsibility of providing human resources for the publishing industry of the future.

A Brief Introduction to Higher Education for the Publishing Industry

The development of higher education for the publishing industry in China is due to the prosperity of the industry itself. Before the 1980s, other than the Beijing Printing College, no other college offered publishing professional education in China. Then in 1983, Wuhan University cooperated with Xinhua Bookstore, which is the largest (and state-owned) book distributor, to set up a department of book distribution identified as the Book and Information Institution of the University. Graduates of this major are awarded a bachelor's degree after four years of study, and more than 2,000 graduates from this department have entered into the publishing business in the last twenty years. After that, some universities such as Anhui University and Nanjing University successively set up majors of book distribution to train distribution professionals. In 1984, Beijing University, Nankai University, and Fudan University set up majors of "editing" which focused on the publishing process. Tsinghua University and University of Science and Technology of China established majors of editing for publishing books in the sciences subject area. There are several colleges offering courses on printing and related graphic arts technologies. The Central Institution of Arts and Crafts and Beijing Printing College have majors of book design and binding.[4] The Beijing Printing College is the largest and most comprehensive site for cultivating professionals for publishing and printing in China. It consists of seven departments, which include publishing, administration, art and design, foreign languages, engineering, electronics, printing, and binding. There are over 2,000 students attending the college, with 700 graduates each year.[5]

The growth of the publishing industry caused demands for more highly educated professionals. Therefore, some universities such as

Wuhan University, Beijing Normal University, Tsinghua University, and Beijing Printing College began to offer postgraduate courses. The graduates receive an MA degree and then teach or enter the publishing and printing professions. However, nearly all postgraduates of publishing studies are required to take courses in journalism offered by their universities. In recent years, teachers and experts in this field have been requesting the establishment of an independent postgraduate program that addresses only publishing studies.

Currently, more than twenty universities offer courses on editing, publishing management, printing, distribution, designing, printing, and binding, with an estimated total of 1,500 to 2,000 graduates each year. However, not all the graduates will enter into the publishing industry; therefore, the actual number of graduates still does not meet the demands of the industry.

After more than ten years of experience, publishing education in universities has formed its own system and character, with its teaching and research levels greatly improved. However, the overall publishing education in China is still in its primary stage and is waiting for further improvement.

Obtaining Employment and Professional Demands

Ten years ago, most students from universities with publishing majors had great difficulties in obtaining jobs in the publishing industry. One reason was that most publishing houses failed to perceive the enormous potential of the students in the increasing competition of not only the publishing industry but also all the other media developing rapidly in China. Recently, these conditions have changed. Many publishing houses have realized that university students with a degree in publishing can adapt themselves more quickly to their work and are more sensitive as well as capable in acquiring titles with market potential. However, publishing houses have become critical in recruiting new employees and in most cases they prefer to hire an MA or Ph.D. from other majors or fields of study. Students who have graduated with a major in publishing should have filled these new entry positions in the industry. As a result, the tough employment situation for publishing graduates in their own profession has not changed very much. On the other hand, they can find employment in more attractive media fields such as newspapers, magazines, and TV stations. However, editing work, in their young idealistic eyes, equals loneliness and boring and tedious work. A professor from the publishing major in Tsinghua University said that most of his students chose to pursue postgraduate studies, study abroad or work in journalism after graduation. Publishing houses were their last choice. A graduate editing major from Beijing Normal University said, "only four out of my 40 classmates went to work for publishing houses this year."

Mid-Level Publishing Education

Mid-level education for the publishing industry includes secondary specialized school and technical school education. The period of schooling of this type usually lasts for two to three years. Mid-level publishing education has the most solid foundation in the entire publishing education system, particularly for graphic arts and printing majors. At present, there are over twenty mid-level education institutions, such as Beijing Printing School, Shanxi Book Distribution School and Liaoning Publishing School in China, providing printing technicians and basic-level managers for the publishing industry. Technical schools mainly assume the responsibility of training printing plant workers and book distributors.

Professional Training

In the publishing industry of China, only a small proportion of the employees have gained formal professional education; most of them need further training by the industry. For more than twenty years, the industry has been continually paying great attention to on-the-job training, which preceded the occurrence of higher education for the industry. Government authorities, such as the Press and Publication Administration of P.R.C. and its local bureaus, mainly sponsor the special or professional continuing training.

Since the beginning of the 1980s, the work force in the publishing industry has experienced speedy expansion. Many new employees, with no former experience in the industry, are in great need of continuous education to improve their professional abilities, skills, and insights. Provincial and Municipal Publishing Bureaus and the Publishing Employees Association at all levels set up short-term, on-the-job-training projects, extending professional knowledge to new employees in the industry. Formal training centers have been established in Beijing, Shanghai, and other cities.

Another driving force has been the wider application of new technologies since 1990. The publishing industry has to become

Table 11.1
A List of Universities That Have Majors for Publishing[6]

Editing Major		
Anhui University	undergraduate	
Beijing Normal University	undergraduate	graduate
Beijing Printing College	undergraduate	graduate
Fudan University	undergraduate	graduate
Henan University	undergraduate	graduate
Nanjing University graduate	two-year program	undergraduate
Nankai University	undergraduate	
Peking University	undergraduate	
Shanghai University	undergraduate	
Shanghai Professional University for employees of publishing industry	two-year program	
Sichuan Academy of Social Science	graduate	
Sichuan University	undergraduate	
Tsinghua University	undergraduate (second Bachelor's Program)	
Wuhan University	undergraduate	graduate
Xi'an Highway University	graduate	
Xi'an Jiaotong University	graduate	
Management of publishing and distribution		
Beijing Institute Of Technology	two-year program	
Beijing Printing College	undergraduate	graduate
Beijing Professional University, East City District	two-year program (night school)	
Nanjing University	undergraduate	graduate
Shanghai Publishing and Printing College	two-year program	
Tsinghua University	certificate program	
Wuhan University	undergraduate	graduate
Book Design, Production and Binding Major		
Beijing Hongqi University of Xuanwu District	certificate program	
Beijing Printing College	undergraduate	
Shanghai Publishing and Printing College	two-year program	
Tsinghua University	undergraduate	
Printing Major		
Beijing Printing College	undergraduate	
Shanghai Publishing and Printing College	two-year program	
Wuhan Technology University of Survey & Mapping	two-year program	
Wuxi University of Light Industry	undergraduate	
Xi'An University of Technology, School of Printing and Packaging Engineering	two-year program	
	undergraduate	graduate
Zhuzhou Institute of Technology	undergraduate	

Table 11.2
Core courses included in the Publishing and Distribution Major offered by Wuhan University[7]

Introduction to Book Distribution, Introduction to Publishing, History of Publishing, Comparative Publishing, Book Marketing, Management of Book Distributing Corporation, Management of Press, International Trade of Books, Electronic Publications, Editing, Storage and Distribution of Books, Demand and Consumption of Books, Introduction to Printing, Book Reviewing, Document Retrieval, Economics of Book Industry, Fundamental Laws for the Book Industry, Publishing Culture, Principles and Application of Computer, Statistics, Accounting for Book Industry, Automation of Books Distribution, Modern Technology of Publishing, Internet and Its Application, Professional English, Practice in Publishing Industry, Graduation Thesis.

more competitive and requires more skilled employees with better computer expertise and abilities. In 1995, the Press and Publication Administration of P.R.C and the National Committee of Education jointly decreed the rules for on-the-job-training eligibility in the publishing industry. The government took the initiative by determining that people such as directors of publishing houses, chief editors, chief editing officers, managers of state-owned printing factories, and the managers of Xinhua Bookstores should be granted certifications on their qualifications for the job before they attain their positions. The Training Center of the General Administration for Press and Publication prepared specific syllabi, contents, and regulations of training programs for different kinds of training and professional studies. Teachers consist of directors of major publishing houses, chief editors, and professors from universities with a publishing major and officials from government authorities. More than 41,600 employees have received training in a total of 795 sessions since 1995.[8] The first rounds of this new training for chief editing officers and high-ranked employees have been carried out. Meanwhile, professional training for editors, proofreaders, and accountants sponsored by local publishing bureaus and the Publishing Employees Association has been completed.

The government has taken other measures to speed up the training of managers for the publishing industry. For example, the Press and Publication Administration of P.R.C. began to implement the "Cross-Century Talent" project in 1996, which aimed at nurturing a large number of senior managers to gain knowledge on publishing issues as well as on economics and business administration skills. Supported by special funds from the publishing industry, selected ambitious young and middle-aged employees pursued further education abroad. The funds also support special research on publishing issues, and award educators and publishing experts who make distinctive contributions to the industry.[9]

The Education Committee of China's Publishers Association is another important source for on-the-job or continuing professional training. The committee has conducted various seminars and short-term training classes since its establishment in 1987. In the 1990s, the committee stepped forward in stimulating on-the-job training and full-time education. From 1993, the committee has held thirty-three sessions for 2,000 employees in the publishing circle.[10] In addition, the committee helps universities such as Tsinghua University, China Science and Technology University to set up publishing-related majors such as editing, publishing, and printing distribution, etc., and arranges for new employees to receive pre-post training in these universities. In cooperation with Beijing Science and Technology University, the committee conducted part-time sessions that lasted for one and a half to two years. The session confers college degrees and is quite popular with employees who do not have adequate professional education.

In addition, some professional organizations such as the China Book Business Report and Open-Book Book Market Research Center conduct training sessions at irregular intervals, focusing on book market research, book advertising and promotions, book distribution, publishing management and strategies, and other topics that seem to be especially relevant in relation to the many changes that are taking place in the market.

The Problems of Education for Publishing and Conditions the Industry Faces

Almost all officials in government publishing institutions, observers in the industry, and teachers in universities and colleges realize that current education for publishing in China falls far behind the varied and rapid development of the industry. In fact, the shortage of qualified personnel has already become a major impediment to the development of the publishing industry.

The problem is quite common in all countries, including European countries and the United States, whose publishing industry is well developed. Regular education for publishing is still underdeveloped in Western countries, since the publishing industry has always been regarded as a low-technology industry generating marginal income. However, rapid economic, cultural, and technical development in recent years has pushed the whole industry to the very front of the development and application of digital technology. Even traditional labor-intensive bookselling, both wholesale or retail, has become a highly technology-intensive industry because of the use of information technology, including online booksellers, which are based on digital technology. E-Book, Print-on-demand, and the Internet are all based on the new high technologies. At the same time, the frequency of mergers and acquisitions increase the intensity of competition in the industry, but are no less than in any other industry. All these change factors make the industry's need for qualified personnel more pressing, and the lack of supply or qualified managers more obvious.

The problem appears especially demanding in China. On one hand, employees in the industry have much to learn and to update. Besides learning new publishing technologies to meet new challenges, publishers in China need to learn the basic ABC's of market economics, such as marketing and financial management, long understood by their Western counterparts. On the other hand, education for publishing is seriously backward. The present education system for publishing was established in the era of planned economy. Due to its old knowledge structure, the education system cannot reflect the newest developments of the world publishing industry, nor can it even cope with the reality of the domestic industry. There is a lack of a group of experienced professional teaching staff equipped to deal with both Internet technology and world economic management. Basically, there is still no professional education and training system with rapid adaptability to meet the fast changing knowledge, skills, and technologies. The present human resources are far behind the industry demand for rapid development, which is highly knowledge and technology intensive with the related organizational, financial and international management skills.[11]

It has become imperative that the publishing industry in China become part of the market economy. Publishing houses are changing from simply production type organizations to a planned national economy to produce books. The industry's expectations for employees have undergone great change. It is no longer expected for editors to be only good at manuscript editing. They are expected to be able to select good works and to understand production, marketing, bookselling, and then the economics of publishing with related financial results. Higher education for publishing currently available is still lacks courses that address these aspects of the present and future industry. Disciplines in publishing are divided among operational streams, such as book or magazine publishing, editorial, printing and distribution (sales), in order to train specified personnel for specific skills with related job functions. This is primarily due to the fact that publishing has long been operated under planned mechanisms and divided clearly with functions in the industry. However, this is unfavorable to the training of general and financial management professionals, as well as to making better use of higher education resources. Another challenge is that the current publishing disciplines focus improperly on editorial courses, with the training of marketing personnel relegated to a level of secondary importance. Until recently, the functions of advertising, promotion, sales and marketing have been identified as "distribution." Meanwhile, the industry is in urgent need of graduates in publishing management that include the skills of marketing and advertising, promotion and publicity, research, and copyright with related sales of international rights. In fact, only the publishing and distribution major of Wuhan University, among the numerous universities and colleges, has always focused on the training of marketing personnel in publishing. At the same time, Beijing Printing College is the only college in the higher educational system to provide English-language courses for publishing. This problem is not to be changed in a short period of time, although the industry has already realized that it has to face the international market as well as competitions from international counterparts.

Most faculty lack practical experience in publishing and are limited to teaching theories. In addition, few universities and colleges have facilities to provide students with opportunities to practice their publishing training for a long period, which makes graduates unfamiliar with the specific publishing practices and fosters a lack of practical working ability. The publishing of teaching materials for the industry is just beginning to take shape. A series of eighteen professional books sponsored by the Press and Publications Administration has already been published, with universities and colleges publishing quite a few other titles. However, the content of current materials is apparently out of date and needs urgent updating. Relevant teaching materials in publishing have not caught up with the new technology, new media, and new management required by the globalization of publishing.

The lack of research and development in the publishing industry is an immediate result of its education system. First, there is no powerful fundamental research group to conduct deeper and thorough studies into the development of the publishing industry both in China and in the world and make a comparison of the two. Second, the publishing industry in China still needs to adapt to a market economy with related financial management. Publishers have long been protected by government policy, provincial territories and government distribution systems. Without competition pressure from foreign capital, the internal or national competition is not yet intensive. That is why the need for specialized management skills is still not urgent and most publishing houses are satisfied with conventional experiences and practices.[12]

Lack of funds is another cause for research in the publishing industry not to be able to go into a deeper and wider range of studies. Higher education in China is still dominated mainly by a planned economy system with funds provided by the state. There is not enough funding to meet the current needs for research and development within the rapidly changing publishing industry as just one segment of the overall communications industry. Publishing houses, while employing graduates of universities with publishing majors, will not invest in higher education. A professor noted, "Publishing houses welcome our graduates, but care nothing about our development." The bookselling and distribution major of Wuhan University was the only program to receive investment from the Xinhua Bookstore at its beginning, while editorial discipline in Nankai University and Tsinghua University received insignificant funding from the Press and Publications Administration, all these being one-time, lump-sum grants. The only way for universities and colleges to obtain funds is to open courses in adult education and training programs for staff in publishing houses. Shortage of funding may be a problem for all institutions of higher education related to other industry-specific programs. However, the potentially high profits of the publishing industry and their benefit from higher education make the indifference of publishing houses even more of a current problem.

The Influence of China's Admission to the WTO

The influence of China's admission to the WTO may be the most interesting topic in the Chinese publishing industry in recent years. According to China and United States' intergovernment agreement on China's admission to the WTO, China has pledged the entrance of foreign capital in the distribution of publications. Such an agreement would allow foreign publishers to distribute tapes, videos, books, and magazines throughout China in forms of joint ventures and cooperative companies. Foreign capital may threaten the distribution industry in China, which may further affect the publishing field. Indeed, the publishing industry in China can hardly match its international counterparts in their ability and experience in market competition, capital, management, application of new technologies, and human resources. It is reasonable to be worried that the entrance of foreign publishers, powerfully equipped with capital, may have a negative impact on the publishing industry in China. Admission to the WTO means that the publishing industry in China will have to face up to the international market, as well as provide new opportunities to further push Chinese publications into the international market. The potential for export sales and sale of translation or co-publishing rights can be realized more effectively in the future.

With the market inadequately developed and capital investments highly restricted, establishing strategic alliances may be an ideal method for the publishing industry in China to improve its competitiveness and ability to withstand risks. At the end of 1999, domestic super bookstores, such as Beijing Book Building, Shanghai Book City, North China Book City, and Shenzhen Book City, established the China Super Bookstore Union and strategic partnership of cooperation. It is their goal to make full use of resources and to resist external pressures. It is also their plan to conform to the developing trends in world bookselling with an emphasis on super-size stores to be run as chains. Since the first publishing group in China was founded in Shanghai in February 1999, five more have been established all over the country as the country's version of "conglomerates" in the Western economy. Several distribution groups have been formed with the Xinhua Bookstore.

The Use of New High Technology

The application of new technology is the most revolutionary event in contemporary publishing industry worldwide. The use of digital and web information has a powerful influence on traditional publishing industry and brings new changes in operations and marketing of the current traditional publishing and printing industry. Multimedia publications and online services threaten to substitute established publications and services, such as the replacement of print on paper publications with CD-ROM, CD-I, and E-Book products, and the replacement of the traditional distribution process with online subscription models or online booksellers and the replacement of the current printing with Print On Demand systems. China has allowed foreign capital to participate as ISP and ICP in the development of the telecommunication market after its admission to the WTO, which means an opening up of online publishing. New technologies have become an area of penetration or opportunities for international capital to take part in the publishing industry in China.[13]

The publishing industry in China is falling far behind the advanced international level in its use of high technology and the development of multi-media publications. Computers are installed in publishing houses and the widespread availability of the Internet provides publishers with a broader view. However, a certain number of managers, directors, editorial and marketing personnel have yet to change from traditional working methods. Paperwork is still being handled by hand in many publishing houses. The role of the PC should not only be considered in relation to sophisticated word processing and desktop publishing systems. There is a real lack of the PC in relation to effective management information systems for operating expenses, sales, cost of sales, inventory and cash flow as part of an overall business system. The overwhelming majority of grassroots bookstores still operate and calculate by hand. The cash register is not a PC with the related and integrated sales, inventory, order processing and cost accounting system. Computers are just decorations for aged editors. However, computer knowledge and skills have become a necessity for young people in the industry. This may be and should be reflected in the courses of universities and colleges mentioned above.

The intensifying competition under the new global conditions demands higher qualities of management and labor force in the publishing industry. The related urgent need of employers to find qualified personnel provides opportunities for further development of publishing education in China. The publishing industry in China, faced with challenges of both WTO membership and new technologies, has become aware of the importance of qualified personnel to cope with the future competition and has focused on steps to train such personnel.

Changes in the focus and arrangement in disciplines and courses clearly indicate shifts in teaching to meet demands of the contemporary industry. The Discipline Catalog of Ordinary Universities and Colleges, published by the Ministry of Education, has combined the original separated disciplines of editorial, design, and production and book distribution into one, namely editorial and publishing management. The combination or coordination of specific skills is an effort to meet the practical needs of the industry, and to better allocate resources in universities and colleges. There is a real initiative to develop an integrated higher education for editorial, publishing, and distribution and to provide highly qualified new publishing personnel with a solid foundation and a wide range of knowledge. Courses of computer skills and the Internet are beginning to be universally established in universities and colleges, while economic courses of marketing are gaining emphasis. The Press and Publication Administration has started its second round of vocational training, shifting focus towards new technologies and new knowledge. Courses of computer knowledge and application have been added to training programs of the Education Committee of the Publishers Association. Reform of teaching materials in universities and colleges has become the order of the day.

The whole scale of education for publishing is expanding. The urgent need of the industry is providing better employment prospects for graduates, while enlistment in publishing disciplines in universities and colleges, such as Beijing Printing College and Beijing Normal University, is increasing. Such changes form sharp contrasts with that of earlier years when it was hard for graduates to get a position in the industry and the whole scale of publishing education was on the decline.

The large publishing organizations made an urgent call for qualified personnel, familiar with the publishing operation and management of modern enterprises and applicable skills of the newest science and technology. The first thirty-three MBAs in publishing graduated this year. People are expecting them to bring better management ideas and effective operational skills back to the publishing houses they originally worked for prior to pursuing the new graduate studies.

Tsinghua University, with majors in science and engineering, is now providing courses and further degrees in publishing to graduates. These graduates are trained in the basics of science and engineering as their original major, as well as in additional systematic knowledge of publishing and editing, both of which are welcomed by publishing houses in science and technology. It is easy to have employees who have gained knowledge in certain disciplines and are familiar with editing and content development. However, the advantage to these graduates with double degrees is that they can combine the two and turn science into subject area productivity in publishing.

The publishing industry in China is speeding up in training qualified personnel capable of international trade in publishing. In April 1999, a training delegation of thirty young managers from publishing houses was sent to participate in the training of international publishing presented by the Center for Publishing of New York University. There have been two more groups since then, each staying for eighteen days, and the sessions are mainly about managing publishing, marketing and distribution as an integrated process.

Exchanges with the foreign publishing industry, which began in the early 1980s, has intensified in recent years. All levels of government authorities intermittently send publishers to visit countries well developed in publishing and to participate in various international book fairs, as well as invite foreign experts to teach in China. The Beijing International Book Fair (BIBF), which has successively been held eight times, is drawing more and more attention from both domestic and international publishing houses. It has become a means to improve mutual understanding through special programs and presentations. Teachers in universities and colleges often visit overseas universities to survey their educational system. New York University's Center for Publishing has developed the closest relationships with the publishing industry in China. However, current engagement with foreign counterparts is still far from satisfying for the industry to learn about international publishing management specifically and new media developments in general. An official in the Press and Publication Administration suggested that current exchange channels are still not enough for information flow, and that it is willing to send personnel to study overseas and to train publishing personnel. The need for continued education and training has been recognized and it is now important to explore how to carry out future programs that will insure both high quality and results for the individuals involved as well as for the publishing industry's growth in China.

Notes

1. "China Statistical Data Collection of Press and Publication in 1999," China Labor and Society Guarantee Press, 2000, pp. 1, 8, 243.
2. "The Book Sales of China in 1999," China Book Business Report, May 28, 2000.
3. Guo Jin, "China Publishing Industry: Take a Deeper Breathe," Publishing Plaza, April 1999, p. 7.
4. Qin Wang, "A Comparative Research in Publishing Education between China and Abroad," China Book Business Report, Beijing, China, April 1999, p 12.
5. www.bigc.edu.cn.
6. Qin Wang, "A Comparative Research in Publishing Education between China and Abroad," China Book Business Report, Beijing, China, April 1999, pp. 18–20.
7. Ibid., p 39.
8. *Press and Publishing Journal* (August 29, 2000).
9. Mili Li, "The Construction of Publishing Major in China," *China Publishing Journal* (August 1999): 18.
10. Statistical data of various professional training courses held by the Education Committee of China's Publishers Association from 1993 to 2000.
11. Youxian Yu, "Speed up the Reformation of Publishing Industry to Meet the Challenge of 21st Century" (January 2000): 28-30.
12. Sanguo Cheng, "The Present and Outlook of Chinese Publishing Industry," China Book Business Report (October 22, 1999).
13. Xin Chen, "China's Publishing Industry Faces the Challenge of WTO," *Wide-view of Publishing* (July 2000): 4-13.

Contributors

Robert E. Baensch is associate professor of publishing and the director of the Center for Publishing, New York University. Before joining NYU in 1996, he was a publishing consultant and president of Baensch International Group Ltd. New York. He was senior vice president for marketing for Rizzoli International Publications, Inc. Prior to that, as director for publishing at the American Institute of Physics from 1988 to 1991, he was responsible for over sixty journals, a book program, database information services and online publishing. From 1983 to 1988, Baensch was vice president-marketing of Macmillan Publishing Company where, in addition to a full range of marketing and sales responsibilities, he directed the Macmillan Software Company and English as a Second Language Multimedia Program. Before 1983, he was president of Springer Verlag New York, and from 1968 to 1980 vice president and director of the International Division of Harper & Row Publishers, Inc. Baensch started his publishing career with the McGraw-Hill Book Company, where he was manager of the Translation Rights Department and editorial director of the International Division.

Baensch has served on the Board of Directors of the Association of American Publishers and chaired their International Division and the Professional/Scholarly Publishing Division. He has also served on the Board of Directors for the Society of Scholarly Publishers. He has served on the Executive Group Council of the STM-International Group of Scientific, Technical & Medical Publishers from 1987 to 1990. He is currently on the Board of Directors of Island Press in Washington D.C., Book Industry Study Group Inc. in New York, and Transaction Publishers in New Jersey. In January 2000, Mr. Baensch was appointed the new editor for the *Publishing Research Quarterly* journal. He is a faculty member of New York University's Center for Publishing and the Stanford University Professional Publishing Course.

Address for communications: Robert E. Baensch, Director, Center for Publishing, New York University, 11 West 42nd Street, Room 400, New York, NY 10036. Email: Robert.Baensch@nyu.edu

Charles Buckwalter is president and founder of Conexion International Limited, Inc. a private consulting firm for the international publishing industry, established 1991 in Miami, Florida, specializing in Asia-Pacific, Latin America, and U.S. Hispanic markets. The company provides consulting/management services for business-to-business or consumer publishers seeking to launch, expand, adapt or improve their publications for international markets. Among his active clients are: MediMedia Pacific Limited, Hong Kong/Singapore, which publishes dental news magazines for China, CIS countries, Latin America and Poland; and Carvajal International S. A., Cali, Colombia, one of Latin America's largest book and magazine printing conglomerates.

Buckwalter's background covers more than thirty years of career experience in corporate staff positions such as international vice president, publisher, marketing director, sales director, and promotion/research manager with well-known publishing firms including Thomas International Publishing (industrial product magazines), Billboard Publishing Inc. (music and entertainment industry), Cahners (trade show division), McGraw-Hill (business and trade magazines), *News World Communications* (daily newspaper), and the *Miami Herald* (daily newspaper).

He is the author of "white papers" for the American Business Media association on publishing in China and in Latin America. He is fluent in English and Spanish, and holds a BS degree in public relations/advertising from Columbia University, New York.

Address for communications: Charles Buckwalter, President, Conexion International Ltd., Post Office Box 835425, Post Office, Miami, FL 33173-4235. Fax: 305-595-4692. Email: conexion@earthlink.net

Hu Shouwen is president and publisher of China Youth Press in Beijing. He started his career as a journalist for seven years and he has conducted research on editing science from the sociological perspective during this time period. He is currently a guest professor in manuscript editing at the Printing Institute of China and at Shanxi Normal University. Hu Shouwen is executive director of China's

Association of Editing and Redaction; vice chairman of the Association for Promotion of International Co-publishing; and author of several papers for the professional editorial sectors of the publishing industry. The State Council awarded him the "Expert with Outstanding Contributions" title. He received a BA degree in philosophy from Shanxi University and a MA degree in art from China's Academy of Art Research.

Address for communications: Hu Shouwen, President and Publisher, China Youth Press, 21 Dongsi Shiertiao North, Beijing 100708, China. Tel: 010-640031803. Email: hsw@cyp.com.ch, www.cyp.com.cn

Li Yuanjun is president of Jieli Publishing House and an experienced publisher in China. She began to work in publishing in 1981, developing and editing TV shows and many books which won the "National Book Award," "Five One Project Award," and "China Book Award." These are the three most important awards in China's book publishing. Over the years, she has developed good publishing management skills, and has worked on building a strong editorial program and exploring international cooperation. She has recognized the importance of actively participating in such industry events as the Bologna Children's Book Fair and the Frankfurt Book Fair. Yuanjun Li has taken an important role in China's children's publishing industry and the recognized imprint of the Jieli Publishing House. Since 1991, she has won many national awards including the National Advanced Worker in Publishing Industry, National Advanced Worker for Children, and the National Best 100 Publishing Worker. She also has won the highest award in China's publishing industry, which is the Taofeng Award.

Address for communications: Ms. Li Yuanjun, President and Publisher, Jieli Publishing House, 9 South Yanhu Road, Nanning 530023, Guangxi, China.

Lin Chenglin is the overseas editor for *China Book Business Report* (CBBR) in Beijing, China. He received an MA in philosophy of science and technology from Jilin University in 1997 and is currently working on a Ph.D. at the same school in the subject area of the philosophy of communications. He has worked for CBBR since 1998, first as an editor for the CBBR *Book Review Weekly*, then as a researcher and reporter on overseas publishing. He was born in 1972 in Changchun, Jilin Province, in northeast China, and lives in Beijing, where he moved in 1998.

Address for correspondence: Lin Chenglin, Overseas Editor, China Book Business Report, Room 2304, FLTRP Building, No. 19 Xisanhuan Beilu, Beijing 100089, PR China. Tel: 86-10-68917695 ext. 2356; Fax: 86-10-68917689; Mobile: 86-13501278372. E-mail: chllin@263.net

Ian McGowan was, until 2002, director of the Center for Publishing Studies at the University of Stirling, United Kingdom, where he ran the postgraduate courses, which were taken by many experienced publishers. He has lectured in many parts of the world and has taken a special interest in publishing in East Asia, having edited a textbook for Chinese publishing professionals. As a member of the Scottish Arts Council, Ian McGowan has chaired its Literature Committee and various panels, with overall responsibility for public policy and national support of authorship and publishing. His publications include books and articles on international publishing and on eighteenth-century literature. He is now a consultant in media and cultural issues for official government departments and commercial organizations.

Address for communications: Dr. Ian McGowan, Center for Publishing Studies, University of Stirling, Stirling FK9 4LA, Scotland. Email: idm1@stir.ac.uk

Sun Qingguo is executive manager of the Beijing OpenBook Market Consulting Center. His main focus is on the research of the rapidly changing book industry of China. He was former vice president of the education publisher of Hebei province and earlier served as president and editor-in-chief of the Hebei Art Publishing organization. He started his career in the world of books by first working in and then serving as president of the Xinhua bookstore. Sun Qingguo also started one of the first online bookstores in China. He has been active in book publishing and distribution for more than thirty years in China.

Address for communications: Mr. Sun Qingguo, Executive Manager, Beijing OpenBook Market Consulting Centre, 17 Fuchengmen Street, 11 floor, Beijing 100037, China. Email: company@openbook.com.cn

Wang Jixiang is now chairman of the board of China Science Publishing and concurrently president of China Science Press. He was selected to chair the board of the China Science Publishing Group in 2000. He was from 1993 to 1996 the vice president of Science Press. Wang Jixiang was president of the Wuhan Branch of the Chinese Academy of Sciences from 1985 to 1993. In January of 2003, he was elected to the National Committee of the Chinese People's Political Consultative Conference. He is also a professor of the Chinese Academy of Sciences.

Address for communications: Wang Jixiang, President, Science Press, 16 Donghuangchenggen North Street, Beijing 100717, China Email: wjx@cspg.net

Yang Deyan is director general of the Commercial Press and from 1991 to 1996, he was director of the Department of Foreign Exchange and Cooperation for the General Administration of Press and Publications in Beijing. Yang Deyan worked for the Chinese Embassy in Germany and Switzerland as first secretary from 1987 to 1991. He started his career in publishing as editor, was promoted to director of the editorial department, and then assistant to the director general of the Commercial Press from 1966 to 1986. He is a graduate of Shanghai Foreign Language University in 1966.

Address for communications: Mr. Yang Deyan, Director General, The Commercial Press, 36 Wang Fu Jing Street, Dajie, Beijing 100710, China. Email: cppost@pubkic2.bta.net.cn

Yu Huiming is currently attending the graduate program of Media Studies at the Newhouse School of Public Communications, Syracuse University, after working for two years as reporter for the China Book Business Report in Beijing, China. He received his BA degree in economics from Nankai University and an MA degree in publishing from the Beijing Normal University. He is the author of three books and has had numerous articles published in newspapers and magazines in China. Huiming Yu is now pursuing research on the effects of media in China as a result of the foreign media groups and organizations entering his country.

Address for communications: Huiming Yu, Yijinfang Street 45, East Room 501, Quzhou City, Zhejiang 324000, China. Email: hyu03@mailbox.syr.edu

Zhang Bohai is chairman of the China Periodicals Association and vice president of the Publishing Association. In that capacity, he has attended several international meetings and presented papers on such topics as "A New Age of Magazine Development in the Asia-Pacific Region," presented in Seoul, Korea in April 2002. He started his publishing career as senior editor and, in 1983, was promoted to deputy editor-in-chief for the People's Literature Publishing House in Beijing. Zhang Bohai was educated in Chinese literature at Shandong University and then joined the faculty there to teach literature.

Address for communications: Zhang Bohai, Chairman of China Periodicals Association, Vice President of the Publishers Association of China, North Building—Room 202, Longii Plaza, Jinbaojie, Chaonei Nanxiaojie, Dongcheng District, Beijing 100005, China. Email: cpazh@public3.pta.net.cn

Index

The index lists subjects, names of persons, organizations mentioned in the text, and authors of publications referred to in the text. Chinese names are listed in the traditional last name first sequence. Initials and acronyms are listed as words; therefore, the "ASEAN" follows "Association of Southeast Asian Nations" or "WTO" follows "World Trade Organization."